House Paints, 1900–1960

History and Use

research in conservation

The Getty Conservation Institute
Los Angeles

House Paints, 1900–1960

History and Use

Harriet A. L. Standeven

The Getty Conservation Institute

Timothy P. Whalen, *Director*
Jeanne Marie Teutonico, *Associate Director, Programs*

The Getty Conservation Institute works internationally to advance conservation practice in the visual arts—broadly interpreted to include objects, collections, architecture, and sites. The Institute serves the conservation community through scientific research, education and training, model field projects, and the dissemination of the results of both its own work and the work of others in the field. In all its endeavors, the GCI focuses on the creation and delivery of knowledge that will benefit the professionals and organizations responsible for the conservation of the world's cultural heritage.

Research in Conservation
The Research in Conservation reference series presents the finding of research conducted by the Getty Conservation Institute and its individual and institutional research partners, as well as state-of-the-art reviews of conservation literature. Each volume covers a topic of current interest to conservators and conservation scientists. Other volumes in the Research in Conservation series include *Stone Conservation: An Overview of Current Research* (Doehne and Price 2010), *Alkoxysilanes and the Consolidation of Stone* (Wheeler 2005), *Analysis of Modern Paints* (Learner 2004), *Effects of Light on Materials in Collections* (Schaeffer 2001), *Inert Gases in the Control of Museum Insect Pests* (Selwitz and Maekawa 1998), *Oxygen-Free Museum Cases* (Maekawa 1998), *Accelerated Aging: Photochemical and Thermal Aspects* (Feller 1994), and *Statistical Analysis in Art Conservation Research* (Reedy and Reedy 1988).

Published by the Getty Conservation Institute
Getty Publications
1200 Getty Center Drive, Suite 500
Los Angeles, California 90049-1682
www.gettypublications.org

Gregory M. Britton, *Publisher*

Tevvy Ball, *Project Editor*
Sheila Berg, *Manuscript Editor*
Pamela Heath, *Production Coordinator*
Hespenheide Design, *Designer*

Library of Congress Cataloging-in-Publication Data
Standeven, Harriet A. L.
House paints, 1900–1960 : history and use / Harriet A.L. Standeven.
p. cm.
Includes bibliographical references and index.
ISBN 978-1-60606-067-4 (pbk.)
1. Paint materials. I. Title.
TP935.S75 2011
667'.6—dc22

2010050176

Printed in Hong Kong

Contents

Foreword

On most occasions, when the Getty Conservation Institute receives requests to publish colleagues' work, there is an immediate and obvious fit to one of our current areas of focus. If I am to be totally honest, Harriet Standeven's research on house paints was not one of them! However, one need not dig very deep to realize that these paints are encountered by conservators in a wide range of areas of twentieth-century cultural heritage—and not just those dealing with painted door frames or architectural finishes. Although Harriet discusses many such areas, perhaps the most surprising is the widespread use of house paints by modern artists, including arguably the most influential of all—Pablo Picasso.

Picasso was one of the first artists to explore the possibilities of these ready-mixed, widely available, and generally affordable paints, and there is every reason to suspect that this choice of nontraditional material was part of his larger quest to challenge all the boundaries of convention in his art. Later on, a great number of important artists turned at some point to house paints, including Willem de Kooning, Francis Picabia, Jackson Pollock, Bridget Riley, David Alfaro Siquieros, and Frank Stella; the list becomes even more impressive when one includes the sculptors who painted their objects with architectural and industrial coatings.

But, of course, with every new type of paint comes a new set of conservation challenges—hence the need for the conservation profession to understand more completely the kinds of materials likely to be found in these commercial paints. The GCI's Modern and Contemporary Art Research Initiative was conceived to address the needs of conservators in dealing with the wide array of materials encountered when treating twentieth-century art, and this book fits squarely within that mission. For conservators to succeed in this challenging endeavor, it is essential for them to understand as comprehensively as possible the factors that govern the behavior, stability, and potential deterioration of the materials in their care.

This volume provides a historical account of the development of the American and British house paint industries in the first half of the twentieth century. We hope that it will prove essential reading for conservation professionals treating objects from this period.

Timothy P. Whalen, Director
The Getty Conservation Institute

Acknowledgments

I would like to thank Stuart Croll, Paula Dredge, and Richard Mulholland for their insightful comments on my text. At the GCI, I am grateful to Tim Whalen for his support of this publication, and to Tom Learner and Cynthia Godlewski for their help and advice in the preparation of this book. I would also like to thank the copy editor, Sheila Berg, and the team at Getty Publications. I am also indebted to the Victoria and Albert Museum, which allowed me time from my duties to write this book. And finally, I would like to thank my husband, John, whose support and encouragement made this project possible.

H.A.L.S.

Introduction

Paints and coatings are so ubiquitous that no part of our lives is unaffected by them: they protect and decorate items as diverse as the cars we drive, the structures we build, the containers in which we store our food, and the homes in which we live. That excellent results can be obtained by an unskilled, inexperienced individual—more often than not painting an ill-prepared surface—is testament to the achievements made by the paint industry with regard to the development of modern, ready-to-use coatings.

Household paints are complex systems that must perform under a range of conditions. An exterior house paint must be easy brushing, self-leveling, elastic, durable, and quick drying. The binder of an interior wall paint must produce a film with uniform hiding, smoothness, and sheen and with maximum resistance to staining and dust collection. It must also be tough enough to withstand the action of various cleaning agents and must not discolor if exposed to heat, light, or shade. Interior trim paints need to be glossy, uniform in color, and hard enough to withstand knocks and scuffs. Last, varnishes for floors, woodwork, or furniture must be hard and tough enough to withstand cleaning and the spillage of food, drink, grease, and other deposits. Of course, no single binder will fulfill all these criteria, and many different types have been used over the years. Today's household paints for interior and exterior use are easy to apply, durable, washable, and fully recoatable, but this was not always the case, and the history of paint technology is really a history of the improvements made in both raw materials and manufacturing processes. Many of these developments were made in Germany in the years before World War II, and while they are mentioned where essential, a detailed discussion of the German chemical industry is outside the scope of this book.

The period under review is the late nineteenth to mid-twentieth century, during which the greatest developments in household paint manufacture were made. From the late nineteenth century we see the first factory-made paints and an increasingly sophisticated range of raw materials available to the manufacturer, and by the mid-twentieth century all the paint binders based on synthetic resins that are still utilized today had been developed.

The versatility of the raw ingredients used in paint and varnish manufacture makes the organization of this book complex: for example, in the early 1950s a flat interior wall paint might comprise distemper, oil emulsion, alkyd emulsion, oil paint, enamel, solvent-borne alkyd, or an emulsion paint based on a synthetic polymer. It therefore makes little sense to organize the content in terms of the end use of the paint. This book is instead structured by binder type and is broadly chronological. It might therefore be necessary for the reader to refer to various chapters in order to gain the complete picture of a certain raw material or paint binder.

Chapter 1 offers an overall history of the developments made in the modern paint industry and discusses the factors that had a significant impact on it: notably the development of ready-mixed paints in the late nineteenth century and the introduction of synthetic resins in the early twentieth. The focus of chapter 2 is traditional solvent-borne paints based on natural oils, gums, and resins. These were the only materials available for paint manufacture until the advent of synthetic resins and were used by the professional housepainter in the production of hand-mixed paints and by the manufacturer in the formulation of ready-mixed oil paints, varnishes, enamels, and gloss paints. Ready-mixed paints based on traditional materials were available from the late nineteenth century until around the 1960s, when they were largely displaced by synthetic alternatives.

Chapter 3 discusses traditional water paints that were used to coat interior and exterior walls. These include simple limewashes and distempers, which have been used for centuries; casein paints, the use of which became widespread in the late 1920s and 1930s; and oil-bound water paints, or oil emulsions, which became available in the late 1930s. For interior applications, distempers, casein paints, oil emulsions, and flat oil paints were in direct competition with one another, until all were largely superseded by emulsion paints based on synthetic polymers, introduced in the mid-twentieth century.

Chapter 4 describes nitrocellulose lacquers. Originally developed as an industrial spray coating for the automobile industry, nitrocellulose was also formulated into quick-drying lacquers for brush and spray application that were available in the United States and the United Kingdom from the mid-1920s. However, problems with the brushing lacquer, including solvent sensitivity and poor brushability, led to its being quickly superseded by alternative products based on natural or synthetic resins combined with drying oils. Although these were still available in the 1940s, by that time they had been largely superseded by the synthetic resins described in chapters 5 and 6.

Chapter 5 covers the synthetic resins that began to replace natural products for the manufacture of oleoresinous finishes in the late 1920s and early 1930s, namely, phenol-formaldehyde and oil-modified alkyds. Phenol-formaldehyde resins were reacted with tung, or combinations of tung and linseed, and were used in the formulation of many of the "four-hour" enamels that were advertised throughout the 1930s. However, the tendency of phenolic resins to discolor rendered them unsuited to pale tints, and it was not until oil-modified alkyds were introduced that a quick-drying, hard, glossy finish with good color became available. Although originally developed in the early 1930s, oil-modified alkyds did not gain widespread use until the late 1940s and early 1950s; they remain one of the most important paint binders today.

Chapter 6 investigates emulsion paints based on the synthetic polymers introduced in the late 1940s and early 1950s. It should be noted that these are called latex paints in the United States and emulsion paints in the United Kingdom. In the United States, the term *latex* was originally applied to styrene-butadiene paints only, but it has since been adopted to describe all waterborne dispersions. For clarity, the term *emulsion* is used throughout this book. Many monomers were available for the production of paints, but the types used in decorative emulsions were styrene-butadiene, which was introduced in the United States in the late 1940s; polyvinyl acetate (PVAc), which was introduced to Britain and the United States also in the late 1940s; plasticized polystyrene emulsion, which was introduced to the U.S. market in the early 1950s; and acrylics, which became available in the United States and Britain in the mid-1950s. Chapter 6 also includes a discussion of the acceptance of synthetic resins by the paint industry in both countries.

Chapter 7 considers the conservation implications of the presence of commercial paints, discussing the weaknesses and potential vulnerability of each binder type.

ARTISTIC USE OF COMMERCIAL PAINTS

The many good properties of commercial paints have ensured that they have been used for a diverse range of applications, and along with protecting and decorating our built heritage, they coat articles such as furniture and toys and are frequently found in works of art. Commercial materials appeal to artists for a number of reasons, including the handling properties and surface characteristics offered, their increased durability, or perhaps for more complex sociopolitical reasons relating to a product's status as an inexpensive, everyday material not normally associated with fine art.[1] For these and many more reasons, artists worldwide have used commercial paints since at least the early decades of the twentieth century, when they became widely available as ready-to-use products. They are present in works of art in the form of both ready-mixed paints and unpigmented resins; the latter might be mixed with other materials to modify properties such as viscosity and drying time, or they may be pigmented by the artist. Although a wide variety of materials have been employed, the product that has the longest history of artistic use, and remains a popular choice today, is ready-mixed household gloss paint.

Gloss Paints: Oleoresinous and Oil-Modified Alkyds

The most ubiquitous commercial paints present in works of art are household gloss, or enamel, paints, and artists have made extensive use of them since good-quality products became widely available in ready-mixed form in the early twentieth century. Gloss paints have many characteristics that appeal to artists: they are not only capable of producing a smooth, glossy finish free from brush marks but also can be poured directly from the can. And, as noted above, their status as inexpensive, everyday materials has been exploited by artists wishing to make a sociopolitical statement.

Pablo Picasso was an early user of household paint; he began to experiment with the French oil-based paint Ripolin in 1912. Ripolin also appears in the 1920s work of Francis Picabia and the 1930s work of Joan Miró. The British artists Christopher Wood and Julian Trevelyan used household gloss paints in the 1930s (Cross 2008), and Ripolin was used extensively by the Australian artist Sidney Nolan from the mid-1940s. Jackson Pollock and Willem de Kooning also began to experiment with gloss paints in the mid-1940s, and by the mid-1950s and early 1960s a new generation of British artists, including Gillian Ayres, Peter Blake, and Patrick Caulfield, was using them extensively. They remain enormously popular to the present day.

Because gloss paints have been so widely used by artists for so many years, it is possible to trace the history of their manufacture through technical analysis of works of art. For example, the gloss paints used by Picabia (plate 1) and Picasso in the first decades of the twentieth century comprise traditional oleoresinous binders, while the paint on Picasso's 1968 painting *Nude Woman with Necklace* (plate 2) consists of an oil-modified alkyd. These developments are also evident in the paintings by the American abstract expressionist Jackson Pollock. He made extensive use of household gloss paints for his poured paintings from the 1940s to the early 1950s, the period during which major changes in formulation occurred.[2]

As might be expected, before 1947 the gloss paints on Pollock's work comprise only traditional oleoresinous binders. For example, *Composition with Pouring II* (1943, Hirshhorn Museum and Sculpture Garden) contains a drying oil with considerable amounts of colophony (rosin): a cheap, friable resin used in the manufacture of inexpensive interior paints. Pollock's 1947 painting *Full Fathom Five* (1947, Museum of Modern Art) also contains traditional oleoresinous house paints, as does Tate's *Summertime* of the same year. However, by 1947 paint manufacturers were making increasing use of alkyd resins in the formulation of gloss paints, as evidenced in *Number 1A* (1948, Museum of Modern Art; plate 3), in which the dripped black paint is oleoresinous but the cream dripped paint is an oil-modified alkyd. By 1949 the poured paints on Pollock's work are almost exclusively oil-modified alkyd.

Pollock's paintings also mirror the technical advances made in alkyd manufacture. For example, his *Number 14* (1951, Tate Modern), contains a phthalic anhydride/glycerol alkyd typical of early gloss paint formulation, but *Yellow Islands* (1952) comprises a phthalic anhydride/pentaerythritol alkyd co-esterfied with a semidrying oil, a binder used in increasing amounts by the early 1950s.[3]

As discussed in chapter 5, alkyd resins were slower to become widely available in the United Kingdom, and correspondingly, a poured painting dating from 1951 by the British artist and designer Jon Catleugh was found to comprise a drying oil, rosin, and glycerol:[4] glycerol-esterified rosin, or ester gum, was a widely used ingredient for the manufacture of inexpensive interior gloss paints at that time, especially in the years after World War II when materials were scarce. Gloss household paints based on oil-modified alkyds continue to be widely used by artists and are frequently detected in works of art.

Nitrocellulose

The Mexican muralist David Alfaro Siqueiros was probably the first artist to use nitrocellulose. Siqueiros's choice of nitrocellulose lacquers for his 1930s murals (see plate 4) was not only underpinned by the need to find durable materials for his outdoor artworks but also manifested his firmly held political beliefs: Siqueiros was a committed Communist, who believed that the only method by which truly revolutionary art could be produced was through the materials and techniques of modern industry.[5] He was one of the first artists to actively seek out synthetic resins and was certainly the first to articulate his choice (Siqueiros [1934] 1975).

Siqueiros's name is firmly associated with DuPont's industrial automobile finish, Duco. He began to use Duco in the 1930s: dripping, pouring, and spraying the material to produce a range of unusual visual effects. Siqueiros taught the use of these materials and techniques in his 1936 New York Experimental Workshop. It was established primarily as an artists' collective to provide posters, floats, and other such propaganda material for the Communists' cause. As the workshop's name suggests, Siqueiros embraced experimentation with new materials and techniques, which he taught to attendees, including José Gutiérrez, Jackson Pollock, and Morris Louis, all of whom experimented with unusual materials and techniques in the years to follow.

Nitrocellulose automobile finishes were also used by the Canadian artists Ted Godwin and Roy Kiyooka for their late 1950s abstract paintings (Leclerc and Barclay 1992: 118, 133), and nitrocellulose spray paints for refinishing automobiles were used by the British artist Richard Hamilton for some of his 1950s pop art paintings (Crook and Learner 1999: 68–83). Nitrocellulose refinishing paints were also used by the American sculptor Donald Judd for many of his 1960s fabricated aluminum sculptures, where they were chosen for the unique optical properties they offered. Nitrocellulose is available today only for very specialist applications, such as the restoration of historic vehicles. Due to its scarcity, it is unlikely to be found on recent works of art, although it was used for the re-restoration of Judd's *Untitled* (1965, Whitney Museum of American Art; plate 5) (Khandekar et al. 2008).

PVAc

PVAc has been used by artists in both solution and emulsion form since these types became available in the early 1930s and mid-1940s, respectively. Solution polymers are dissolved in organic solvents, dry by evaporation, and remain soluble in their carrier solvents. Emulsion polymers, in contrast, are dispersed in water and dry by evaporation of the water, followed by coalescence of the polymer particles. Once dry, emulsion polymers become insoluble in solvents, although they are swelled by most solvents, including water.

Solutions of vinyl acetate were used by Gutiérrez and Siqueiros from the early 1930s (Gutiérrez 1956), and the American artists Lue

Osborne and Cordray Simmons used PVAc solutions in 1938 (Lodge 1988). What is likely solution PVAc was mixed with water-based ink by the American sculptor David Smith in drawings dating from the late 1950s and early 1960s.[6]

The German-born American painter Karl Zerbe was among the first to experiment with PVAc emulsions, in 1945, reportedly using Borden Company's Polyco 953-7A emulsion (Lodge 1988) (plate 6). And in the same year the American artist Alfred Duca collaborated with Borden to develop artists' Polymer Tempera paints, which were based on PVAc emulsions.[7] However, these artists' paints met with limited success due to problems with pigment dispersal. Around 1957 the Australian artist Sidney Nolan began to use PVAc emulsion, which he mixed with dry pigments to make an inexpensive, very-quick-drying paint (Crook and Learner 1999: 22).

To date, household emulsions based on PVAc resins are infrequently found in works of art. But the 1960s op art paintings of the British artist Bridget Riley are among the most well-known examples of their use (plate 7). Riley liked emulsion paint because of the very flat, impersonal surface it produced (Crook and Learner 1999: 143). However, that the paints are PVAc based is coincidental rather than a conscious choice by Riley: PVAc was and still is used almost to the exclusion of other binders in the manufacture of standard interior emulsion paints in the United Kingdom. Household emulsion paints were used by the American artist Robert Ryman for his *Varese Wall* (1975, Dia Art Foundation, New York) (Rivenc et al. 2010). The work is repainted when deemed necessary and thus contains many layers. Analysis revealed that the original layer contained styrene, whereas the uppermost layer comprised PVAc; both were used extensively in the formulation of interior house paints in the United States.

Acrylic

The occurrence of acrylic is far more prevalent than PVAc: indeed, it is the synthetic resin that occurs with most frequency in paintings belonging to the Tate, London (Learner 2008: 6). However, acrylics are usually present in the form of artists' quality paints or acrylic gesso primers, and it is rarer to find in either household paints or unpigmented emulsions in works of art.

The first solution acrylic (Acryloid) was introduced by Rohm & Haas in 1927. This early resin was based on soft polymethyl acrylate and required baking at high temperatures to achieve full gloss and hardness, so it was an industrial material. Solution acrylics do not appear to have been used by artists until the 1950s, by which time they were based on harder poly-*n*-butyl methacrylate. These solution resins were manufactured by Rohm & Haas (Acryloid F-10; Paraloid F-10 in the United Kingdom) and DuPont (Lucite 44 and later Elvacite 2004) (Borchardt-Hume 2008: 238). Lucite 44 was certainly used by the Canadian artist Jock Macdonald in the mid-1950s: instead of adding turpentine or extra oil to his oil paints, Macdonald used Lucite, which increased both the fluidity and the drying time (Leclerc and Barclay 1992: 210). Poly-*n*-butyl methacrylate also formed the basis of Bocour's Magna artists' paints, acrylic solution paints that were introduced in 1949 and used extensively by artists such as Morris Louis, Roy Lichtenstein, Mark Rothko, and Willem de Kooning during the 1950s and 1960s (Crook and Learner 1999: 27).

Unpigmented acrylic emulsions were also acquired by artists and used as a binding medium for their paints. For example, a number of Canadian abstract artists working in the 1950s and 1960s, including Douglas Morton, Art McKay, and Oscar Cahén, used Rohm & Haas's Rhoplex AC33 and AC34. The artists reportedly obtained their resins from local paint manufacturers (Leclerc and Barclay 1992: 214). Despite these notable exceptions, the use of commercial acrylics is limited among artists, and if acrylic is encountered in artworks, it usually originates from artists' quality paints.

Without conducting scientific analysis, it is not possible to ascertain the binding medium present, and although it is recorded that the Canadian abstract artists Tom Hodgson and Jean Goguen used proprietary emulsion paints in the late 1950s (Leclerc and Barclay 1992: 214), their composition is unknown. They might be PVAc, acrylic-vinyl acetate comonomer, or even styrene-butadiene, as all three types of resin were produced in Canada at that time.

Traditional Water Paints

Although there is little evidence to suggest that traditional water-based paints have been used by artists in the manufacture of portable art, they were certainly used to decorate walls and buildings (see chapter 3). For example, many English monastic wall paintings dating from the twelfth to fourteenth centuries reportedly comprised pigment, slaked lime, and milk (Mulholland 2010: 155), a coating equivalent to fortified distemper and a forerunner to casein paints. And although casein itself has been widely used by artists, either in commercial tubed form or as a raw ingredient for mixing paints, there is at present little evidence to suggest that artists made use of household paints based on casein.[8]

Phenol-formaldehyde

There is thus far no evidence of artists using household paints based on phenol-formaldehyde resins, although the resin has been found in works of art. Phenol-formaldehyde was certainly used by Rothko for his Seagram murals (late 1950s to early 1960s, Tate Modern, London; Kawamura Memorial Museum, Japan; and National Gallery of Art, Washington, D.C.) and was detected on the murals in the Tate's collection. Although the source of phenol-formaldehyde in this instance is unclear, it is likely to have originated from a printing ink. Phenol-formaldehyde was widely used for this application in the late 1950s, and the almost ubiquitous presence of lithol red (PR:49) on the Seagram murals (Carlyle et al. 2008), a pigment commonly used for printing, suggests that Rothko may have obtained his materials from the printing industry.

A second notable occurrence of the presence of phenol-formaldehyde is in the faked Vermeer paintings executed by Hans van Meegeren in the

1930s. In this instance, its detection provided incontrovertible proof that the paintings were indeed faked. Van Meegeren chose phenol-formaldehyde, as it enabled him to re-create the visual effect of a cracked seventeenth-century paint film. He bought the resin as a thick solution in benzene or alcohol, which he diluted with the essential oils of lavender or lilac. He then pigmented the resin and heat-hardened the finished painting for four to six hours at temperatures of 120°C to 130°C to create the craquelure that is characteristic of aged paint (Breek and Froentjes 1975).

Mixed Materials

Artists who choose commercial products tend to experiment with a number of them, and many artists are notable for their prolific use of a range of commercial products. For example, David Smith used PVAc, alkyds, and nitrocellulose lacquers alongside artists' materials such as Magna, acrylics, drawing inks, and casein temperas during the 1950s and 1960s (Mulholland 2010). Rothko's use of commercial materials is similarly diverse, and the Tate's Seagram murals alone contain oil-modified alkyd, acrylic (*n*-butyl methacrylate), phenol-formaldehyde, dammar, egg, and oil (Carlyle et al. 2008). The materials chosen by Robert Ryman are equally wide ranging; the eighteen paintings belonging to the Dia Art Foundation and the Solomon R. Guggenheim Museum in New York were found to contain at least eleven different binding media, including PVAc, styrene, oil-modified alkyd, and acrylic (Rivenc et al. 2010).

THE ROLE OF ANALYSIS

The scientific analysis of artworks not only helps in the detection of forgeries but also is an important tool for the accurate identification of materials present in works of art. This ensures that artworks receive appropriate care and treatment; in addition, it informs our understanding of artistic practice and technique and can clarify misunderstandings surrounding the particular materials that an artist used. A tendency for commercial materials in works of art to be referred to by certain trade names when cataloged by museums is widespread and can result in erroneous assumptions about the particular material present. For example, a number of Sidney Nolan's paintings from 1943 and after are described by the Art Gallery of New South Wales, Sydney, as "synthetic polymer paint," due to a mistaken belief that the Ripolin household paint Nolan used was based on an alkyd binder at that time. In fact, Ripolin was not put onto an alkyd system until the mid-1950s.[9] But the case that illustrates most clearly the confusion that can result from a reliance on brand names is that of Jackson Pollock and DuPont's Duco.

The relationship between Pollock and Duco was established by several means: cans of Duco were photographed in Pollock's studio; Pollock's wife, Lee Krasner, referred specifically to Duco when asked about Pollock's choice of materials; and Pollock certainly would have encountered Duco at Siqueiros's 1936 New York Experimental Workshop (Lake, Ordonez, and Schilling 2004). Although the brand name Duco is

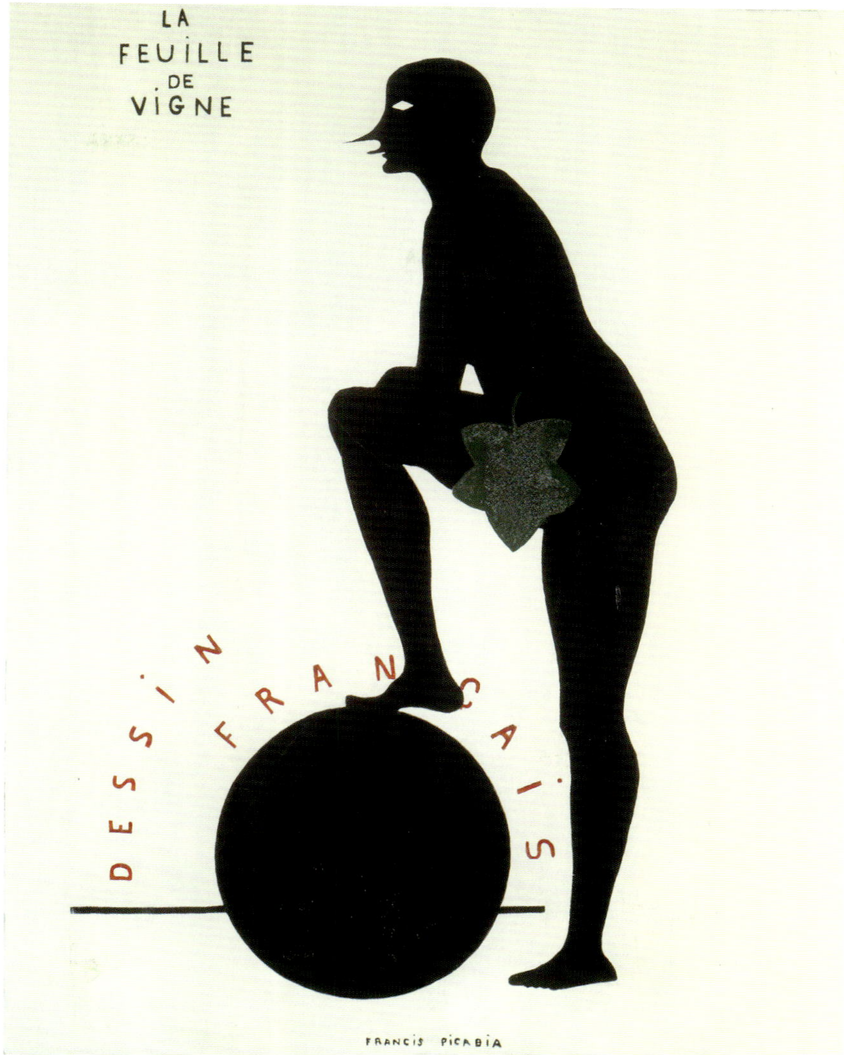

Plate 1
The Fig-Leaf (La Feuille de vigne), 1922
Francis Picabia (1879–1953)
Oil on canvas (traditional oleoresinous house paint)
2062 × 1663 × 70 mm

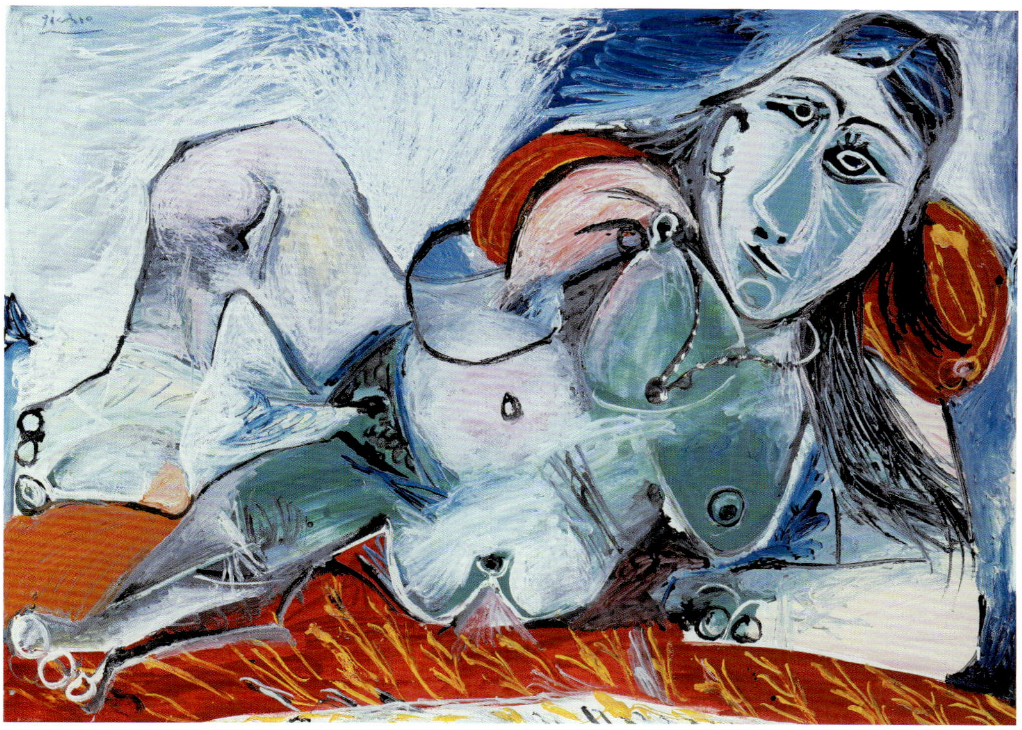

Plate 2
Nude Woman with Necklace (Femme nue au collier), 1968
Pablo Picasso (1881–1973)
Oil and alkyd house paint on canvas
1181 × 1666 x 62 mm

Plate 3
Number 1A, 1948
Jackson Pollock (1912–1956)
Oil and enamel on unprimed canvas (oleoresinous and alkyd enamel paints)
68 × 104 in. (172.7 × 264.2 cm)
Museum of Modern Art, New York

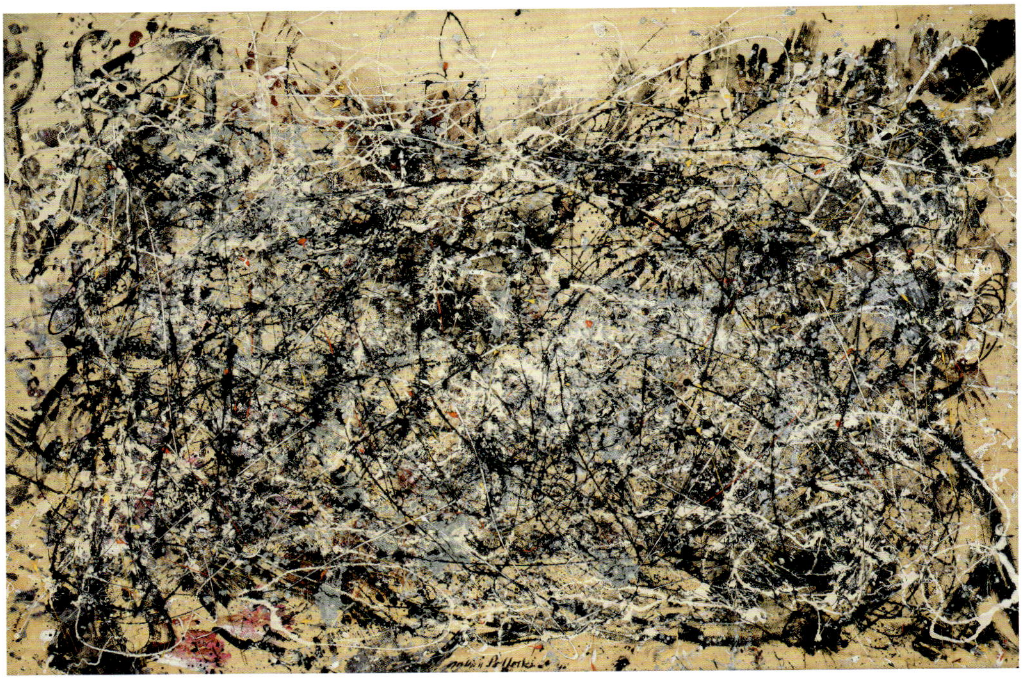

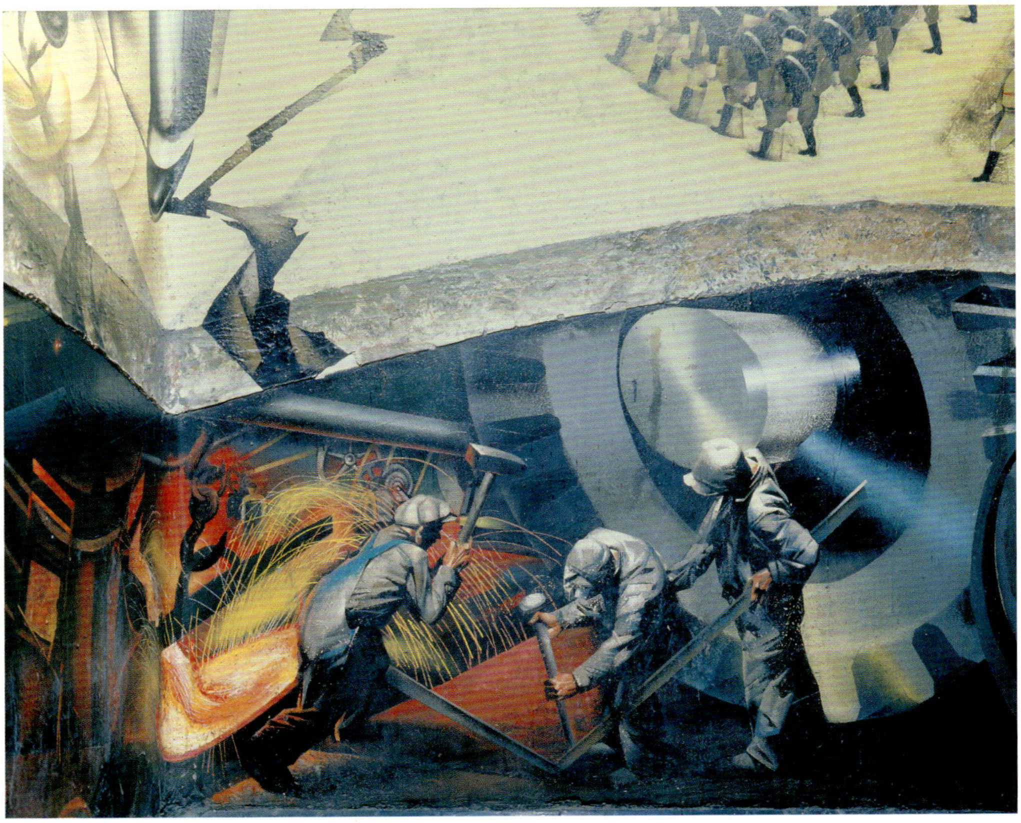

Plate 4
Portrait of the Bourgeoisie: Foundry Underneath a Concrete Slab, 1939
David Alfaro Siqueiros (1896–1974)
Mural. Duco nitrocellulose lacquer on concrete.
 Sindacato Mexicano de Electicistas, Mexico City, D.F., Mexico

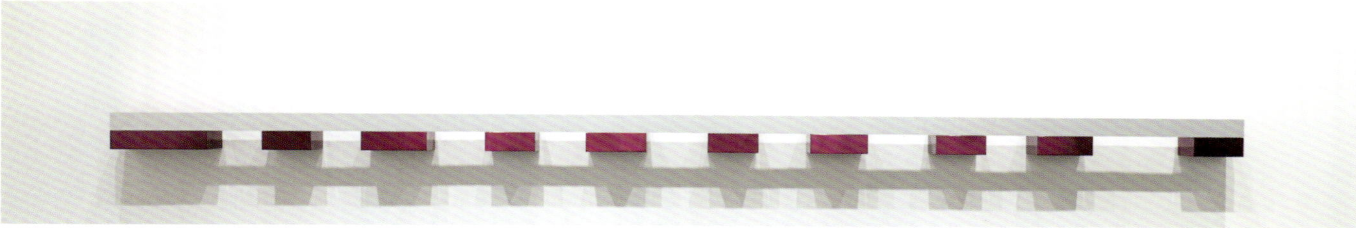

Plate 5
Untitled, 1965
Donald Judd (1928–1994)
Aluminum and nitrocellulose lacquer
8¼ x 253 x 8¼ in.
(21 x 642.6 x 21 cm)
Whitney Museum of American Art, New York; purchase, with funds from the Howard and Jean Lipman Foundation, Inc. 66.53 Art © Judd Foundation. Licensed by VAGA, New York, NY.

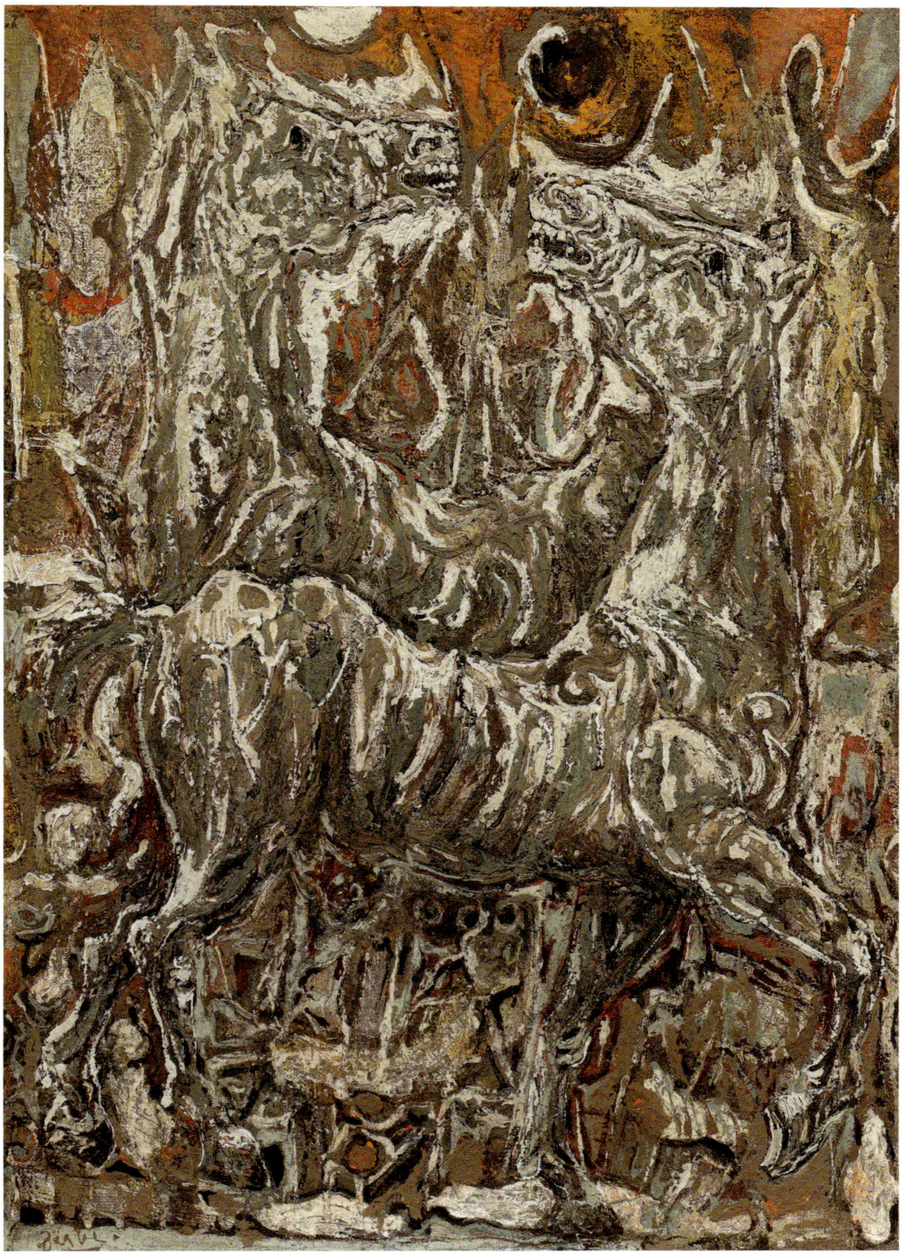

Plate 6
Two Equal Horses, 1950
Karl L. Zerbe (1903–1972)
24 × 17¾ in. (61 × 45.1 cm)
Polymer tempera on canvas mounted on Masonite
Museum of Fine Arts, Boston
Photograph © 2011 Museum of Fine Arts, Boston
Courtesy Mercury Gallery

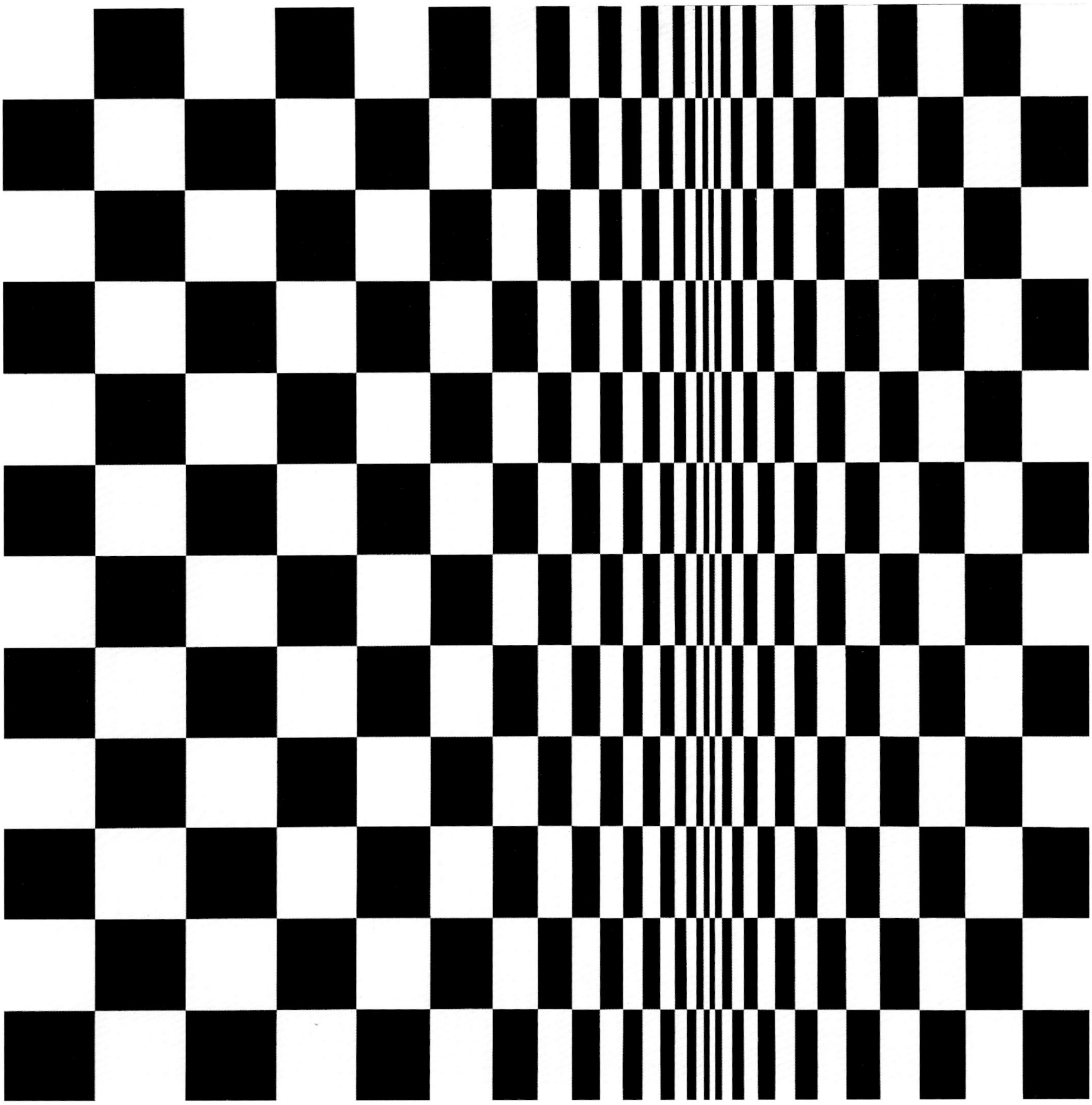

Plate 7

Movement in Squares, 1961

Bridget Riley (1931–)

PVA emulsion on canvas

2172 × 2172 × 40 mm

British Council

Courtesy Karsten Schubert, London

most firmly associated with DuPont's industrial automobile finish based on nitrocellulose, the huge success of this coating had in fact prompted DuPont to exploit the brand, and the company formulated a quick-drying interior gloss brushing lacquer under the same name.

The first version of "Brush" Duco, which was introduced in 1926, was based on alkyd-modified nitrocellulose. However, the extremely poor performance of brushing Duco, including its solvent sensitivity, unpleasant odor, and a drying time so rapid that it was impossible to brush out, rendered it largely useless, and its formulation was swiftly changed to an alternative quick-drying oleoresinous medium. In the mid-1930s brushing Duco was reformulated with an oil-modified alkyd resin (Zintl 1947: 79), and if it was manufactured at all during World War II, its formulation would have reverted to an oleoresinous or similar product, as the use of synthetic resins for the manufacture of decorative paints ceased during that period. By the late 1940s, when synthetic resins again became available for the manufacture of retail paints, brushing Duco would have been reformulated once more with an oil-modified alkyd resin. Therefore, it is likely that the Duco used by Pollock was not the industrial version based on nitrocellulose but a retail gloss paint marketed to the home user. This is certainly borne out by the analysis that has been conducted on Pollock's paintings, as nitrocellulose is seldom found in his work. Although nitrocellulose was detected in the orange paint on the Tate's *Yellow Islands* (1952) (Learner 2008: 10), the nine Pollock paintings dating from 1943 to 1950, examined by Lake, Ordonez, and Schilling in 2004, found no evidence of nitrocellulose, only oleoresinous and oil-modified alkyd paints.

***

The presence of commercial materials can present unique conservation challenges for those charged with the care of artifacts containing them. This book provides a historical account of the developments made in the household paint industries in the United States and the United Kingdom in the first half of the twentieth century. It is intended for any professional caring for heritage that may contain house paints.

Chapter 1

History of the Paint Industry

Today the household paint industry is dominated by a small number of global firms. Paint production techniques are standardized, and the raw materials and technologies used are well established, making many manufacturers' products indistinguishable. The resulting decorative paints are inexpensive to buy and easy to apply. However, before the mid-twentieth century it was a quite different story, and the period from the late nineteenth to early twentieth century arguably saw the most rapid developments in paint and varnish manufacture. Two primary factors contributed to this revolution: the introduction of ready-mixed paints in the mid- to late nineteenth century, and the development of synthetic resins in the first decades of the twentieth.

Prior to the introduction of synthetic resins, protective and decorative coatings were based on naturally occurring oils, gums, resins, waxes, and proteins. The simplest and cheapest comprised limewashes and distempers; more expensive oil paints were reserved for painting wood and metalwork and for applications where greater durability was required. The coating was hand-mixed just before use, and it was common to add extra ingredients in order to improve its properties: limewash might be fortified with tallow to increase its water resistance and weathering; oils and proteins were a frequent addition to distempers to increase their durability; and bodied oils or varnishes might be added to oil paints in order to improve water resistance, gloss, drying time, and leveling.

However, hand mixing is an inexact science, and the huge variation in the quality of raw materials made any sort of quality control or consistency impossible. But this somewhat haphazard method of paint production was all that was available until the late eighteenth century, when the beginnings of the industrial revolution marked the emergence of the modern paint industry. The effects of the industrial revolution on the industry were twofold: not only did newly built bridges, railways, trains, and buildings require protective and decorative coatings, but the increased demands made on the industry led to the beginnings of mass production techniques for paint manufacture. In response to this demand, the first specialist paint and varnish makers emerged in the late eighteenth and early nineteenth century.

By the early nineteenth century the market consisted of many hundreds of small firms, each fighting for orders in a fragmented marketplace. There was no network of roads and canals to facilitate movement of the finished product, so paint companies tended to serve their local regions. Paint formulation was an inexact science, and the jealously guarded recipes were handed down from one generation to the next. But as specialist firms emerged and as production moved from the workplace to the factory, the process for manufacturing paint became increasingly mechanized. Despite this increased professionalism, the industry was still largely seasonal, with many employees laid off over the fall and winter months when the demand for paint decreased. This trend continued for a considerable time, and one large British manufacturer reported that as late as the 1930s if the varnish maker went on holiday, production would cease until he returned ("Story of Paints Division" n.d.: 9). However, an increasing awareness of the importance of laborsaving and quality control techniques can be inferred by the numerous applications for patents concerning manufacturing processes of pigments and paints that were taken out during this period.

READY-MIXED OIL PAINTS

The development of ready-mixed paints can be divided into three phases: (1) the professional painter bought his raw materials individually and mixed them himself before use; (2) semiprepared paints were introduced, which required the addition of oil, driers, or thinners before use; and (3) ready-to-use paints, which were applied straight from the can, became available.

A fledgling industry in ready-mixed oil paints emerged in the mid-nineteenth century, but the paints were so poorly formulated that the pigment often settled into a hard mass at the bottom of the can (Armitage 1955: 229). Not only were manufacturers ignorant of the additives and extenders necessary to ensure the paint's stability, but, in order to cut costs, adulteration with cheap ingredients was widespread: barytes was used in place of expensive pigment, water was added to the binder, and part of the oil might be replaced with cheap substitutes such as kerosene (Heckel 1928: 26). Although additives such as extender pigments were in fact essential to ensure stability and good rheological properties, early manufacturers did not understand how they should be used, and additions were made purely to cut costs. The quality of early mixed paints varied considerably, and it was reported that many manufacturers even put two or three brands on the market: only the best displayed the manufacturer's name and address, and the poorer grades bore a fancy, often fictitious name (Uebele 1913). Although the faults in these early paints rendered them largely useless, their introduction did mark an important milestone: paints could be acquired and used by the amateur for the first time.

The first serviceable ready-mixed oil paints were introduced in the late nineteenth century in both the United States and the United Kingdom. They were available in flat and gloss finishes and for interior and exterior use. However, the use of ready-mixed paints did not become widespread until the 1920s, and although by that time there were many good-quality paints available, their reputation was plagued by the prevalence of poor-quality paints, which made many house painters and contractors reluctant to use them at all. But by the mid-1920s significant improvements had been made, and increasing references to them appear in the trade literature: whether the professional liked them or not (the rise of the home user was sure to affect his business), ready-mixed paints were becoming impossible to ignore.

It is evident that ready-mixed paints were introduced earlier and their use was more widespread in the United States than in the United Kingdom—a situation no doubt brought about by the far greater number of home users of paint in the former country. In 1926 it was claimed that since the 1880s the consumption of ready-mixed paints in the United States had grown to sixty million gallons per year.[1] In 1927 the United Kingdom similarly enjoyed an "extraordinary increase" in the use of ready-mixed paints over the previous year (Cayley 1927: 744). The rapid rise in the popularity of ready-mixed paints can be attributed both to improvements made in raw materials and to an increased understanding of the scientific principles of paint production, as manufacturers recognized the contribution of the physical condition of the raw materials to the durability of the final film.

Although by the mid-1920s there were many high-quality ready-mixed oil paints on the market, a number of factors hindered their acceptance and use: first, the existence of low-quality paints had tarnished the reputation of all ready-mixed paints, and poor-quality products continued to be manufactured in great quantity; second, the raw materials used were so far removed from those that the painter bought that he no longer understood their properties; and third, the painter tended to mix a bit of extra driers or thinners to ready-to-use paints, which would not tolerate adulteration. The result was that many professional painters were mistrustful of all ready-mixed paints and preferred to continue mixing their own from tried and trusted raw materials. Their belief that ready-mixed paints were of inferior quality was certainly justified in many instances: the problem of poor-quality ready-mixed paints occupied many column inches in trade journals throughout the first decades of the twentieth century, and there is evidence to suggest that some British terms of contract dating from the early 1920s specifically forbade their use ("Should high-class decorators" 1922: 14–15).

An issue that was related to quality but was a little more complex was that of adulteration, the substitution of high-quality materials by those of poor quality. Many paints were indeed adulterated with cheap ingredients, but it was in fact necessary to add numerous components to ensure good stability in the can and flow upon application. Although pigment extenders (e.g., silicates of alumina, magnesia, and various calcium carbonates) were essential components of the paint system, it was difficult to convince the house painter that they were not simply cheap bulking materials. However, it is evident that even by the late 1920s manufacturers were formulating paint without fully understanding how the raw materials might interact, as evidenced by one painter's recollections: "One may open a can of light blue and discover it to be a sickly green; or a can of pea green, and find it a muddy olive." Such changes were caused by interactions between incompatible materials (Kelly 1927: 795).

But the professional painter himself must take much of the blame for the poor performance of some ready-mixed paints: he found it difficult to relieve himself of the habit of "improving" the paints by adding a little extra turps, oil, or driers and in the process inadvertently adulterating them to such a degree that they became useless (Raffé 1928: 219). Carefully formulated ready-to-use enamel paints were especially vulnerable to this, and the painter should perhaps have heeded this warning: "Finally *never* under any circumstances add anything to an enamel or enamel paint before use. It is the paint makers' business to make paint and the decorators' business to put it on, and neither can usurp the provenance of the other" ("Some points" 1905: 29–30).

Painters were also warned against mixing paints to obtain a particular shade unless the manufacturer specifically stated that the paints were fully intermixable, as dark and pale gloss paints were often put on different binders. For example, high gloss in tints tended to be achieved by the introduction of a resin varnish, whereas high gloss in whites was achieved with a treated oil binder. If the acid number of the varnish was high, it could "feed," or thicken, a paint containing basic pigments, such as the zinc oxide found in whites, resulting in an unworkable paint (Snelling 1952a: 52). Manufacturers therefore produced a range of tinting colors for their paints.

Despite these drawbacks, there were of course so many advantages to using ready-mixed paint that as the manufacturers improved them, increasing numbers of people used them. Along with their ease of use, a major advantage was the ease with which colors could be matched. In previous times, when painters hand-mixed their materials, getting a good color match between batches could be truly challenging. Ready-mixed paints made this the manufacturer's problem. Furthermore, considerable labor costs could be saved by the use of ready-mixed paints: the time taken to physically mix the materials was removed, and the use of manufacturers' binders with greater body meant that fewer coats were required.

The late nineteenth century also saw the introduction of the first ready-mixed water-based paints, and by the first decades of the twentieth century ready-mixed oil and water paints were formulated for all purposes, including interior and exterior woodwork, interior walls, and masonry, available in flat, gloss, and eggshell finishes. By the mid-twentieth century the materials used in these paints were arguably at the most diverse: synthetic resins had been introduced but had not completely displaced natural products, and wartime shortages of raw materials had forced companies to look for alternatives. Shortages of linseed oil encouraged research into bodied oils and oil-in-water emulsions, as both techniques utilized the available oil most efficiently. And it was also at that time that alternative oils such as tung, dehydrated castor, perilla, soybean, tall oil, and fish were introduced. The impact of the rise of the petrochemical industry was also significant, as this industry made available to manufacturers a huge range of solvents, monomers, and other ingredients that became staples of the modern industry. But it was the introduction of synthetic resins in the 1920s, 1930s, and 1940s that had the most dramatic impact on the paint industry.

DEVELOPMENT OF SYNTHETIC RESINS

The synthetic resins used by the paint industry were developed by the plastics industry. Although the pioneering work into polymerization techniques was conducted in the late nineteenth century, this research did not come to fruition until the 1920s and 1930s. The initial impetus for research into synthetic polymers was conducted in order to find substances to replace the naturally occurring gums and resins that were used as molding materials and electrical insulators. The discovery of methods by which synthetic polymers could be manufactured was concurrent with the development of the commercial production of solvents such as alcohols, esters, and ketones, which paved the way for the manufacture of an astonishingly diverse range of specialized plastics and coatings, which was to revolutionize both industries.

Work into the development of synthetic resins began in the mid-nineteenth century, and it was J. J. Berzelius who, in 1833, coined the term *polymer* to describe the precipitated mass left after his experiments in the formulation of new compounds. Little could be done to utilize this discovery, however, until an understanding of the processes governing the reaction were better understood. It was not until the early part of the twentieth century, largely due to the work of two research scientists, Wallace Carothers in the United States and Hermann Staudinger in Germany, that chemists were furnished with a better understanding of molecule functionality, polymerization, and the fundamental difference between thermoplastic and thermosetting resins (Elliott 1981: 13). Once the basic mechanisms governing the behavior of synthetic polymers were better understood, research into their applications for technology began in earnest, and it is from the early part of the twentieth century that numerous influential products began to appear. The introduction of paints based on synthetic and semisynthetic resins in the first decades of the twentieth century had a dramatic impact on the paint industry: from the solvent-borne nitrocellulose lacquers introduced in the mid-1920s, phenol-formaldehydes in the late 1920s, and oil-modified alkyds in the 1930s to the waterborne emulsions based on vinyl and acrylic monomers in the late 1940s and mid-1950s, respectively. Concurrent with this were developments in other essential components, in particular, pigments.

KEY DEVELOPMENTS IN PIGMENTATION

Although the field of pigments is vast and lies outside the scope of this book, it is useful to discuss some of the more important developments. Pigments provide opacity and color to a coating, and they might also protect the substrate or the binder from the deleterious effects of ultraviolet radiation. The interaction between the pigment particles and the paint medium affects the film's rheological properties and its durability. The type of pigments present can also affect the rate at which the paint film dries; for example, the presence of metal salts tends to increase the rate of drying.

Until the mid-nineteenth century many of the commonly used pigments comprised highly toxic raw ingredients such as lead, arsenic, and mercury. One of the most important of these was lead white, which had been the only good-quality white pigment available to manufacturers until the latter part of the eighteenth century. It was largely in response to the toxicity of lead that zinc oxide, an alternative white pigment, was introduced. This pigment became available commercially at the end of the eighteenth century (Boxall 1978a: 24), from which time it was widely used in paint and varnish manufacture: lead and zinc were often used in combination for both interior and exterior finishes.

The next significant development in white pigments was the introduction of the first titanium dioxide pigments in the mid-1920s. Titanium dioxide was originally available in anatase form only. The anatase form has good opacity and high tinting strength and chalks slightly to retain a clean, bright surface.[2] However, its chalking rate was somewhat uncontrolled, and it was not until the far superior rutile form of titanium dioxide was introduced in the early 1940s that the very high tinting, opaque, and durable pigment that we know today became available. The two forms were frequently used in combination to achieve a controlled rate of chalking, which was desirable on exterior paints in particular at that time.

The pigmentation of interior paints was improved considerably by the introduction of lithopone in the 1880s. Lithopone comprises varying proportions of zinc sulfide and barium sulfate, with a small amount of zinc oxide. Lithopone has excellent covering power and was widely used to pigment both oil- and water-based flat wall paints ("Flat Paints" 1931: 663). However, its use was restricted to interior applications only, as it turned gray if exposed to sunlight.

The early nineteenth century saw the development of many new colored pigments, including synthetic ultramarine, zinc chromate, lead chromate, cadmium sulfide, chromium oxide, and emerald green (a complex copper acetate-arsenite). By 1845 about fifty different pigments were available for use in British paint manufacture (Boxall 1978a: 24). In the late nineteenth century aniline and azo dyes were introduced, many of which, however, had poor lightfastness. Metallic phthalocyanines (organo-metallic pigments) such as phthalocyanine blue were also developed in the 1920s. Colored pigments, especially those based on organic dyes, tended not to be lightfast, and it was not until the 1950s that lightfast yellow, red, maroon, and orange pigments were developed. Today there are many hundreds of pigments available to the paint manufacturer, of both natural and synthetic origin.[3]

Chapter 2

Traditional Oil-Based Paints

Prior to the introduction of synthetic resins in the first decades of the twentieth century, all solvent-borne paints and varnishes were based on naturally occurring oils, gums, and resins. Linseed was by far the most important drying oil, and it was used in vast quantities by both the professional house painter, who hand-mixed his own paints, and the manufacturer, who supplied him with semiprepared and ready-mixed paints. The late nineteenth and early twentieth century saw the development of many new manufacturers' materials, including the rosin derivative ester gum, tung, perilla, soybean, dehydrated castor, fish, and tall oils.

Paints and varnishes based on natural products formed the basis of a range of oil paints, enamels, gloss paints, and varnishes. The quality of these finishes varied tremendously and ranged from tough, durable exterior coatings comprising high-grade oils and varnishes to cheap pigmented spirit varnishes at the lowest end of the interior paint market. Indeed, many manufacturers made a firm distinction between interior and exterior paints, as each system had specific requirements.

The paint that protects the exterior of wooden houses must be flexible and durable in order to withstand extreme weathering. Tough, elastic binders made of pure oil were therefore used for this application. However, although very durable, paints based on pure oil binders were too soft for interior use, where good weathering was of secondary importance to quick drying and the ability to withstand knocks, scuffs, and the action of cleaning agents. Quick drying and hardness were obtained by the addition of varnishes or natural resins. Although natural resins improved gloss, hardness, and drying time, they had a negative effect on durability, as they contributed to the discoloration, embrittlement, and eventual breakdown of the film.

Although generally excluded from exterior finishes because of their poor weathering properties, natural resins were included in some exterior finishes, where the advantages they conferred—quick drying and good leveling and gloss—outweighed the disadvantage of a shortened life for the film. Natural resin varnishes were thus often included in exterior trim and enamel paints, although the resins were by necessity high-grade types. By contrast, durability was not necessarily a requirement of interior finishes, and a great number of soft, friable, and acidic resins were used in their formulation: this makes the comprehension of the materials used in the formulation of interior paints far more complex than that of exterior paints. Finally, although many manufacturers made a firm distinction between paints for interior and exterior use, there were an equal number that marketed gloss and enamel paints suitable for both applications: these were presumably high-quality finishes that would withstand weathering. One final point worth noting is that although gloss and enamel paints are today associated with the decoration of wood and metal, prior to the introduction of scrubbable and durable interior wall paints based on synthetic resins, they were also used for finishing walls, especially where water resistance and good sanitation were requirements.

DRYING OILS

Drying oils are capable of drying on exposure to air, and they form the basis of all traditional oil-based decorative paints and varnishes, including alkyds (see chapter 5). Oil-based paints and varnishes dry by oxidative polymerization. The polymerization reaction is complex but basically consists of a two-step process: initial uptake of atmospheric oxygen by the double bonds within the fatty acids, followed by decomposition of the oxidation products, which initiates cross-linking. The process of cross-linking forms an insoluble polymer and continues long after the coating has dried, eventually causing brittleness, cracking, and the breakdown of the film.[1]

In order for an oil to dry, a proportion of the fatty acids that make up the constituent triglycerides must be unsaturated. Unsaturated fatty acids contain carbon-carbon double bonds, which provide a site for oxidation to occur. Oils containing conjugated double bonds (e.g., tung) dry far more rapidly than those containing nonconjugated double bonds (e.g., linseed). The presence of certain metal ions, including cobalt, manganese, and lead, increases the drying time of these oils considerably, and oil-soluble compounds of them, such as napthenates, resinates, or their salts, are often added to commercial paints and varnishes to increase drying time (Mills and White 1994: 38). The drying speed and viscosity of oils can be increased by pretreatment such as blowing or boiling: these processes are discussed in detail below.

Linseed Oil

Linseed oil has been used as a binding medium for artistic and decorative paints for hundreds of years, and it remains an essential paint component to the present day. Linseed oil derives from the seed kernels of the flax plant and comprises a complex compound of glycerol and fatty acids. It is an unsaturated oil with excellent drying properties and good flexibility, although it discolors on aging. Linseed oil can be used in its raw, refined, or bodied state, and it was common to use a combination of raw and bodied oil in the formulation of household paints. Bodied oils have increased viscosity and are subdivided into two categories: boiled or blown. Boiled oils have been heat-treated, and the resulting increase in viscosity is brought about by polymerization, whereas blown oils are exposed to oxygen, and the increased viscosity is due to oxidation.

The best grade of oil for the manufacture of artists' paints is obtained by cold-pressing flax seeds, followed by settling or tanking to remove much of the mucilaginous portions, or "foots"; this results in 97% to 98% pure oil. The oil can be further refined through an acid- or alkali-refining process, which removes the remaining foots. Alkali-refined oil is pale in color and practically neutral and is used extensively in the formulation of paints and varnishes. Acid-refined oil is used as a binder to a much lesser extent and found its greatest role as a grinding medium for white lead pigment (Damitz, Murphy, and Mattiello 1943: 199). Raw and refined oils were used extensively in binders for exterior paints; their low viscosity, easy brushing, durability, and flexibility rendered them ideal for this application.

Treatment of Linseed and Other Drying Oils

Boiled oils comprise refined oils that have been heated to high temperatures in the presence of driers. Heating the oil causes it to partially polymerize across the carbon-carbon double bonds, which increases viscosity and drying time, while the driers catalyze the oxidation process. Depending on the treatment carried out, the oils are classed as either dark or pale. Traditionally, lead driers were used, but the subsequent darkening of the oil rendered them unsuitable for use in pale tints, and improvements were made by using lead in combination with manganese oxide. Even paler oils could be made by using soluble resinates, linoleates, and napthenate driers (Damitz, Murphy, and Mattiello 1943: 199). Extra-pale and extra-quick-drying oils could be produced by heating refined oil to a moderate temperature, rather than boiling, and including soluble resinate driers; however, these oils lacked body unless artificially bodied by blowing with air. Although the weathering properties of blown oils are good, they generally imparted too much softness to a paint film to be of much use to the decorator ("Linseed Oil" 1949: 53).

"Stand oils" are thick, viscous oils that are slow drying but have great durability and weather-resisting properties. Historically, stand oil was made by leaving linseed to stand in glass jars exposed to sunlight for considerable lengths of time. More recently, it has been made by heating refined oil to over 300°C in the absence of oxygen. This causes the oil to partially polymerize across the carbon-carbon double bonds. Stand oil has a thick consistency and varies in color. Its excellent durability led to its use in the manufacture of high-grade enamels and paints ("Linseed Oil" 1949: 53).

Use of Linseed Oil

Raw and refined linseed oils found extensive use in the formulation of exterior paints. Paints based on them have low viscosity, are easy brushing, and have good flexibility and durability. They are slow drying, so driers in the form of metal soaps are added to speed up this process. They have relatively low gloss and exhibit poor leveling, which means brush marks remain clearly visible. Refined linseed oil has better color and color retention than raw, so it was used in the best-quality paints and in pale tints. Raw linseed oil binders were heavily pigmented with white lead or lead/zinc combinations, as the bodying of the binder was achieved through pigmentation alone. Although it was frequently used as the sole binder, raw and refined linseed was often used in combination with dark or pale bodied or boiled oils.

Boiled oils are heavier, darker in color, and more viscous than raw oil. The initial dark color of dark boiled oil and its tendency to discolor further on aging rendered it unsuitable for white paints, and the elasticity that all boiled oils impart to a film made it inappropriate for use in undercoats or interior paints. Boiled oil was not often used as the sole binder, although it was recommended as a first coat on stucco or plaster, as it is nonpenetrating and less affected by alkalis than other oils and gives a heavy, elastic coat. A high proportion of boiled oil was also added to roof paints, as it set quickly and was tough and weather-resistant ("Characteristics of Boiled Oil" 1925: 908).

Boiled oil was primarily used as an addition to exterior paints, where it was added to the order of one-third boiled oil to two-thirds raw oil or mixing varnish. The addition of boiled oil increased the gloss, hardness, drying time, water resistance, and body of the coating.

It is evident that opinion was divided over the use of boiled oils: some reportedly used them extensively for exterior work; others did not use them at all ("The Use of Boiled Linseed Oil" 1924: 1001–2). A major drawback of boiled oil was the fact that although the initial gloss was good, it diminished rapidly. By the mid-1930s a writer for the British journal *The Decorator* reported that it was not much used by decorators at that time, primarily because of its tendency for rapid loss of gloss ("Raw and Boiled Oil" 1937: 80). Pale boiled oil was very pale and was suitable for white and pale tints and zinc paints. Because they were more viscous, bodied oils did not penetrate to the same degree as raw oils; they also had better flow and, because they tended to be applied thickly, better self-leveling properties ("Bodied Oil" 1945: 31).

Stand oil was primarily a manufacturers' medium. It was incorporated into slow-drying white enamels and elastic enamel paints. The film was somewhat soft, so it was often combined with a varnish to increase the hardness. Enamels made with first-class stand oil had great durability and weather-resisting properties ("Grades of Drying Oils" 1924: 76). However, finishes made from it were sticky, and brush application was slow and tiring (Snelling 1952c: 61). On the Continent it was reportedly bought by the professional painter along with raw and boiled linseed oil and used as a paint ingredient, although in the United Kingdom it remained a manufacturers' medium ("Grades of Drying Oils" 1924: 76); its availability to the professional painter in the United States is unclear.

Each type of treated oil has certain properties that made it suited to certain applications. Bodied oils have better wetting properties, more flow, and better drying and color retention than unbodied oils. However, they exhibit inferior brushing, and the higher the degree of polymerization, the poorer the brushing. Blown oils have significantly better wetting properties than bodied oils, which in turn are better than unbodied oils.

However, very good wetting properties also results in a tendency to sag on application. Moreover, blown oils are not always compatible with the remainder of the binder, resulting in poor package stability. It was recommended that no more than 25% of binder solids should consist of blown oil (Troutman 1943: 348). Bodied perilla, linseed, and soybean also promoted good flow.

Tung Oil

The introduction of tung oil to the paint manufacturers' stock of raw materials was a significant step in the development of fast-drying, waterproof paints and varnishes, and it was used extensively for this purpose after its widespread availability in the 1920s. Tung was used in its bodied state, either alone or in admixture with linseed oil, and in combination with rosin, ester gum, and phenolic resins. Tung was introduced in the United States in 1896, to Germany in 1897, and to the United Kingdom shortly after (Jordan 1929: 60). Although it was available at the end of the nineteenth century, its use did not become widespread until World War I, when rationing forced companies to look for alternatives to linseed oil. Tung in fact had a number of advantages over linseed: it was cheaper and faster drying, and its water resistance was considerably better. The latter point was especially pertinent, as water-resistant varnishes were desperately needed for airplanes at that time. The "four-hour" and "half-time" enamels that were widely available throughout the 1930s contained tung, often in combination with ester gum or phenolic resin (see plate 8).

Tung oil derives from the nut of the tung tree. It is native to China (hence its alternate name, China wood oil), from where it was initially imported. Tung is a highly unsaturated oil, making it extremely fast drying. Because it dries to a much harder film than linseed, the addition of a resin was not always necessary. However, if resins were required, the superior hardness of tung meant that expensive fossil resins such as copal could be replaced with cheaper, soft resins such as rosin or ester gum, without detriment to the final film (Jennings 1925: 723); in fact, the use of hard resins could cause the oil to gel (Jolly 1930). Early on, manufacturers realized that tung combined with limed or hardened rosin (calcium or zinc resinate) or ester gum imparted superior chemical and water resistance to the final film, and it thus found widespread use in applications where good weathering was required (Taylor 1948: 91), including floor varnishes, marine paint, and exterior varnishes. Tung oil was also much cheaper than either linseed or the fossil resins that had been used in enamel manufacture, and it found extensive use in lower-priced enamels (Uebele 1913).

Varnishes made from tung oil and modified rosin had many other good properties, and a 1924 article describes them as being good bodied, lustrous, durable, and exceedingly pale—pale enough to be used as a wallpaper varnish (*The Decorator,* July 22, 1924). The excellent water resistance of coatings containing tung oil meant that it was also widely used in the manufacture of washable flat wall enamels, as it was scrubbable within forty-eight hours of application (Jennings 1925: 723) and did not impart any gloss to the film; this was due to its natural matte finish and the presence of heavy petroleum thinners that were used to dilute it (Uebele 1913).

The use of tung oil was widespread in paint and varnish manufacture in both the United States and the United Kingdom. The London-based firm Thomas Parsons & Sons claims to have been one of the first firms to use tung oil in place of linseed during World War I (Thomas Parsons & Sons n.d.). By early 1922 it was estimated that 75% of all architectural varnishes produced in Britain were made wholly or partly with tung oil (Bohannon 1922: 237). And by the late 1920s it was reported that 90% of high-grade varnishes sold in the United States contained it (Heckel 1928: 305). In the United States, DuPont carried out extensive research into its use as a component of interior decorative enamels in the late 1920s, although work was halted once the potential of oil-modified alkyds was recognized (Coolidge 1929).

However, the fact that many of these coatings contained tung oil was not widely advertised by the manufacturer for a number of reasons. First, the oil could not be used in its raw state and underwent considerable processing, and companies who had achieved this did not wish to share their knowledge (Coolidge 1929). Second, many of these coatings contained rosin. Rosin is a soft, friable resin that lacks durability and discolors on aging. It was quite rightly considered a low-grade material, and although paint and varnish makers knew that good coatings could be made from it if it were properly processed, they were reluctant to share this information with their consumers (Fasig 1926: 54). Third, decorators were wary of the fact that tung dried so quickly, which was perhaps understandable. To achieve quick drying in coatings based on linseed oil, either lots of drier or lots of resin was required, both of which resulted in embrittlement, cracking, and early deterioration of the film. With tung, however, the quick-drying nature was due entirely to the inherent properties of the oil, and it took decorators some time to fully believe that a film could be so quick to dry and yet still durable. Fourth, decorators were unfamiliar with tung oil and its properties: it required processing before use and was not available for the decorator to buy and mix with his own paints.

One of the main drawbacks of tung from a manufacturer's perspective was the fact that both Europe and the United States were dependent on imports from China, and its supply was unreliable. Furthermore, Chinese tung was processed by means of roasting and pressing, which resulted in dark-colored oils that were unsuited to pale tints. It was also common for Chinese-imported oil to be adulterated with soybean, rapeseed, tea seed, perilla, groundnut, and sesame oils (Jordan 1929), and even grease and fish and mineral oils. Tung was sold by weight, and there were reports of packages of Chinese tung containing sand, earth, and even water to increase its mass (Uebele 1913).

Soon after World War I, the United States, realizing that tung was becoming a staple of paint manufacture, grew reluctant to be wholly dependent on Chinese imports, especially following the dramatic price increases that occurred in 1923. In response to this, the American Tung

Oil Corporation (ATOC) was founded in late 1923, with the expressed aim of establishing a tung crop in the United States. The first seedlings were planted in Florida, Cuba, and Hawaii and elsewhere in the Pacific in spring 1924. In 1928 the ATOC provided manufacturers with its first batch of tung oil for testing. U.S. tung was extracted under clean and proper conditions, and the oil was pronounced far superior to that imported from China; the first large-scale mill came into operation in January 1929 (Jordan 1929).

Despite efforts by the United States to produce a tung crop, by the late 1930s demand far outstripped supply, and it was partly in response to this shortage that alternative oils were developed. Although research into the chemistry underpinning the treatment of oils had begun in earnest during World War I, it did not really come to fruition until the 1930s, and it is from this decade that most of the alternative oils became widely available.

Oiticica Oil

The mid-1930s saw the introduction of oiticica oil from Brazil. It came into use to alleviate shortages of tung, most of which was still imported from China. Oiticica has characteristics and properties similar to tung and was used initially by the U.S. paint industry to make varnishes with properties similar to those containing tung (*The Decorator,* December 15, 1934, 20). It is faster drying than linseed and perilla but slower drying than tung. By the mid-1940s oiticica was replacing a portion of the tung oil in many varnish formulations, and it could be combined with some phenolic resins to produce varnishes with satisfactory exterior durability (Damitz, Murphy, and Mattiello 1943: 200). By the early 1940s oiticica was being used in increasing amounts in the United States due to the shortage of tung, as Chinese imports of the latter had ceased by that time ("Oiticica Oil" 1942).

Perilla Oil

Perilla oil derives from the leaves of *Lamiacea,* a member of the mint family. Perilla is chemically similar to linseed oil. It is unsatisfactory in its raw state due to its tendency to crawl, but heat bodying eliminates this defect. Perilla can be refined and used in a manner similar to linseed. Perilla made up a large proportion of the oils used in U.S. paint and varnish manufacture before World War II but became scarce in the 1940s when imports from the Far East had stopped (Damitz, Murphy, and Mattiello 1943: 200).

Fish Oils

Fish oils used in coatings derive from sardine, menhaden, and pilchard. The crude oil is refined by refrigeration, which removes the saturated fatty acids that hinder drying and cause clouding and discoloration of the oil. Although their drying properties are not as good as alternatives, fish oils were used in combination with other oils, especially if low cost was a deciding factor: for example, in 1953 one U.S. company advertised Crystols, which was based on alkali-refined fish oil and could be added to a range of interior flats, glosses, and enamels and to exterior paints to prevent sagging.[2] Generally, fish oils did not find their way into high-quality paints, and they were never popular, largely due to their odor—whether real or imagined.

Dehydrated Castor Oil

Although pure castor oil will not dry, the removal of a water molecule from its structure results in a drying oil with excellent properties. Castor beans were imported from South America and to some extent from India (U.S. National Research Council 1940: 230). Although primarily used in combination with alkyd resins, dehydrated castor oil (DCO) was also used in traditional oleoresinous binders. DCO became important commercially in the late 1930s, when a shortage of tung oil prompted its commercial development in the United States and Germany. At the beginning of World War II excellent castor oil was available in the United States. By the early 1950s the major use of DCO was in combination with alkyds for the manufacture of interior whites, white baking enamels, and can coatings. The potential was recognized for other applications, including exterior house paints. So, although originally developed as a replacement for tung, by the early 1950s it had become a versatile drying oil with a wide range of applications. It imparted good color-retention, drying, flexibility, and chemical resistance ("Terril Addresses Vehicle Manufacturers" 1953).

SEMIDRYING OILS

Soybean Oil

Soybean is a pale-colored semidrying oil that is used extensively in paint and varnish manufacture. Although its expense precluded wide availability in Britain until the early 1960s, it had been widely available in the United States since it was first imported from China in 1907 (Heckel 1928: 306). A cheap and plentiful supply was subsequently produced, as soya bean grows well in the U.S. Midwest, and from that time it was experimented with in the formulation of paints and varnishes. However, early attempts to formulate paints with soybean and similar indigenous semi- and nondrying oils such as cottonseed and corn had ended in failure (Heckel 1928: 461): they are insufficiently unsaturated to form a hard film in air-drying coatings. However, they were cheap, and although up to 10% soybean could be added to linseed without detection or a compromise of drying properties, the early U.S. varnish industry was replete with attempts to replace large amounts of linseed with soybean, resulting in coatings with poor drying properties, surface tackiness, dirt retention, and lack of durability (Bradley 1951: 854).

Soybean oil was not considered suitable as a high-grade paint ingredient until the 1930s, when its poor drying properties had been overcome by proper processing. By 1934, ten million pounds of the oil was reportedly used in U.S. paint manufacture, and by 1936 there was evidence that Germany was purchasing soybeans from Manchukuo to use in paint production ("Soybean Oil" 1936: 47). During the mid-1930s it was not used as the sole oil in paint binders but in equal proportions with linseed. It remains an important oil in the formulation of decorative alkyds today.

Safflower Oil

Safflower became commercially available in 1949, from which time it was used in paints, varnishes, printing inks, and linoleum manufacture. It has very low linoleic acid content, so coatings based on it have good color and exhibit very little discoloration on aging, even under extreme conditions. It was used in admixture with drying oils and also in the formulation of alkyd resins (Christensen 1950). Safflower oil confers fast drying and good color and color retention, gloss, and water and alkali resistance (Prane 1954).

Tall Oil

Tall oil and its derivatives are widely used in industry and form the basis of adhesives, coatings, cements, asphalts, soaps, and emulsifying agents. Tall oil is a by-product of the Kraft wood pulp process of paper manufacture that turns pine into pulp and paper. The crude oil is a black, oily residue that comprises rosins, fatty acids, sterols, high molecular alcohols, and alkyl chain products. The crude oil is distilled and refined into pitch, and oils that comprise virtually all rosin acids and virtually all fatty acids are produced, depending on end use (Brushwell 1954). It is the tall oil fatty acids (TOFA) (i.e., oleic acid) that are used in the manufacture of paints and varnishes.

TOFA were not used in paint and varnish manufacture until World War II, when shortages of fats and oils encouraged chemists to look for alternative sources of these products. They have a number of benefits over acids derived from vegetable origin. They contain few highly unsaturated fatty acids and thus have good oxidative stability: this confers good color and viscosity stability to coatings containing it. Because it is a by-product of the paper industry, fatty acids are inexpensive, and their price is less likely to fluctuate than oils deriving from crops. Finally, the acids are available in free fatty acid form, unlike vegetable oils, which need to be hydrolyzed in order to release the fatty acids.

STYRENATED OILS

Although initial experiments with the polymerization of oil and vinyl compounds had been investigated in Germany and the United States before World War II, it was not until the late 1940s that the research came to fruition. In the United Kingdom, Lewis Berger & Sons conducted research into the process and in 1948 introduced a binder that comprised styrene polymerized in the presence of oil. The resulting oil represented a considerable savings in linseed, which was still in short supply after the war, and, it was claimed, had better lightfastness and water-resisting properties ("Notes from Manufacturers" 1948: 99). It was recommended for use in enamels and flat wall paints and could be incorporated into alkyd finishes also. At a similar time in the United States, Dow introduced Styrene N-99 modified oils that were intended for trim paints. They were marketed as one of the most durable and cheap paint materials on the market and were fast drying and resistant to yellowing, offered bright whites, and had good durability and water resistance. They were to be added to drying oils and alkyds and were intended for all interior uses, including woodwork and walls.[3]

MIXTURES OF OILS

Traditional household paints contained linseed oil alone, but as an increasing array of oils became available to the paint manufacturer throughout the twentieth century, it became common to use oils in admixture to obtain optimum results. For example, tung oil was frequently combined with linseed in the formulation of quick-drying paints, and blends of soybean and tung (soywood) were also common. Soywood blends, which became available in the United States around 1949, took advantage of the good properties and low price of both oils and could be used as a pound-for-pound replacement for heat-bodied linseed oil. Soywood oil was marketed as suitable for one-coat enamels and house paint formulations (*American Paint Journal,* April 4, 1949, 39). In the same year, Spencer Kellog & Sons advertised its dicyclopentadiene oils, which consisted of linseed and soybean oils reacted with dicyclopentadiene. The resulting oil was fast drying, alkali resistant, completely miscible with bodied or unbodied oils, and recommended for use in varnishes and enamels (*American Paint Journal,* May 30, 1949, 18–19). Admixtures of oils present a considerable problem for those analyzing paints, as it can be difficult to separate and identify the component oils.

VARNISHES

Varnishes are clear, unpigmented films that are used to protect and decorate woodwork, floors, decking, and wallpaper. They have also been added to oil paints to increase gloss, hardness, and drying time, and they form the basis of many ready-mixed gloss paints. Varnishes can be divided into two main groups: oil varnishes that contain hard resins fused with oil and dry by oxidative polymerization, and spirit varnishes that consist of soft resins dissolved in a solvent and dry by evaporation of the solvent. Oil varnishes are further divided into short, long, and medium oil lengths. The oil length refers to the ratio of resin to oil: the longer the oil length, the greater the amount of oil and the more flexible and durable the coating. Increasing the resin content improves the drying time, gloss, and hardness of the varnish, although at the expense of its durability. Different types of varnishes were used for different applications: spirit varnishes might be considered adequate for high-gloss, low-cost interior applications where durability was not an issue, but good-quality long-oil varnishes were employed for exterior applications or floor varnishes, where durability was essential.

Traditional oil varnishes were made by fusing certain gums and resins with drying oils. The resin was heated, or run, in a metal container and the oil added to it; turpentine was also added if required. The varnish was then left in tanks for as long as two years to mature and to allow the mucilaginous portions, or foots, to settle out. This long tanking period was a necessity until synthetic resins were introduced in the early twentieth century. Different types and quantities of oils, gums, and resins produced a range of varnishes for different purposes. The quality of the final product depended entirely on the quality of the constituent ingredients.

By the mid-1930s flat or matte varnishes were also in popular demand. Matte varnishes consisted of an ordinary oil varnish that incorporated waxy material, and sometimes a transparent pigment such as alumina, which had a rough surface texture and a different refractive index from oil, both of which help to reduce gloss. However, it is evident that there were problems with such finishes: matting agents imparted softness to the film, and the finish would not tolerate any adulteration, or the delicate balance of materials would be disrupted (Symons 1933). Varnishes formed the basis of most of the flat and glossy enamels and paints that were introduced in the late nineteenth century, discussed in detail below.

NATURAL RESINS

Resins are an important addition to paints and varnishes, as they increase the gloss, hardness, and drying time. Before the introduction of synthetic resins (see chapters 4–6), they derived from natural sources. Most natural resins consist of terpenoids, which are made up from units of the C5 compound isoprene, which can be formed into many different chemical structures and is also the main constituent of natural rubber (see Natural Rubber in Household Paints, below). The number of carbon atoms in the molecule dictates the properties of the terpenoid, and those found in natural resins are mono- (C10), sesqui- (C15), di- (C20), and tri- (C30) terpenoids. Although one or more are often found in the same resin (e.g., gum elemi comprises a mixture of mono- and sesquiterpenoids), di- and triterpenoids are not found together in the same resin, which is useful for the purposes of chemical analysis (Mills and White 1987: 95). The natural resins commonly used in paint and varnish manufacture comprise di- and triterpenoids.

Generally, triterpenoids are nonpolymerizing but easily oxidizable, which accounts for their poor durability and poor color. Two common triterpenoid resins are the soft resins mastic and dammar. Dammar derives from the family of trees Dipterocarpaceae and is used in the formulation of spirit varnishes. Although films containing it are lustrous and have good adhesion, they are soft, lack durability, and discolor on aging. Dammar has found extensive use as an artists' picture varnish, but its lack of durability made it unsuitable for all but a few specialist applications in commercial paint and varnish manufacture; for example, it was used as a grinding medium for pigments, and varnishes containing it were added to enamels and enamel-type paints. Dewaxed dammar was added to early nitrocellulose finishes to increase adhesion, gloss, and build (see chapter 4). Another soft resin, mastic, derives from the genus Pistacia; its use in commercial paint and varnish manufacture is very limited, so it is not discussed here.

Diterpenoid resins derive from exudations from various trees and plants that are found throughout the world. Most, including the Coniferae, Leguminosae, and Anacardiaceae families produce hard, amber-forming resins, whereas the conifer subspecies Pinaceae produce soft, non-amber-forming balsams. Balsams include common turpentine, rosin/colophony, Venice turpentine, and Canada balsam; the amber-forming resins include sandarac, kauri gum, and the copals. Copal was an essential diterpenoid resin for the formulation of quality varnishes and paints. Copals comprise a wide range of hard fossil and tree resins, which, as a group, vary in hardness, acidity, and solubility. The highest grade for paint and varnish manufacture was Congo copal, a fossil resin that derives from the tree *Copaifera demsusi* (Gettens and Stout 1966: 15). Other copals include Sierra Leone copal, Manila copal, and kauri gum. The amber-forming diterpenoid resins are classed as hard resins, which must be heated to attain solubility in oils and solvents.

Rosin and turpentine were also essential paint ingredients. Both derive from the sticky balsam of the Pinaceae species. The balsam is collected and distilled: the distillate is turpentine, and the sticky residue left behind is rosin. Rosin is also obtained from tall oil. Rosin was used in the formulation of spirit varnishes but has poor color, lacks durability, and is highly acidic. However, the primary component of rosin is abietic acid, a monobasic acid that reacts with metals to form soaps and with alcohols to form esters: both products were used extensively in paint and varnish manufacture, in the form of resinates (oil-soluble driers) and ester gum (Sanderson 1934: 711).

Shellac differs from the above resins in that it derives from an insect rather than a plant source. Shellac is a soft resin that comprises a resinous excretion from the lac insect. It is used for spirit varnishes and makes a good wood sealant, but its moisture resistance is poor. In terms of its incorporation into commercial paints and varnishes, it was used for low-cost interior finishes and also as an addition to the nitrocellulose finishes discussed in chapter 4.

ROSIN AND ROSIN DERIVATIVES

Rosin is a soft resin obtained from the sap of pine trees. When compared to copals and other fossil resins, rosin is inferior, yet excellent varnishes could be made from it if it were properly treated. In its raw state, rosin is too acidic, soft, and sticky to be used in quality paints and varnishes, and it causes livering if used in combination with basic pigments such as lead carbonate, lead sulfate, and zinc oxide. Numerous decorators had experienced firsthand the poor quality of coatings based on linseed and untreated rosin, manifest as rapid deterioration of the film. Indeed, rosin had such a poor reputation that until 1914 there was hardly a paint or varnish specification that did not forbid its use (Jennings 1925).

Despite its serious faults, rosin had been included as a minor ingredient in varnishes for hundreds of years. Varnish makers realized that placing a small amount of rosin in the bottom of the kettle during manufacture prevented the fossil resin from charring and also improved the solubility of the fossil resin in the oil. Rosin was far cheaper than fossil resins, so it was perhaps inevitable that varnish makers replaced increasing amounts of the fossil resin with rosin—with disastrous results. Although possessing a lustrous gloss, varnishes containing significant amounts of rosin were soft and sticky, with very poor water resistance and durability. Rosin thus became inextricably linked with poor quality in the mind of the consumer (Sanderson 1934).

Varnish makers realized that a harder, more durable product could be made by partially neutralizing the rosin with lime (calcium hydroxide), to form calcium resinate. The lime constituted just 6% to 8% of the rosin, as the resinate was rendered insoluble in the oils and solvents used in varnish manufacture if it had been completely neutralized. Limed rosin dissolved in naphtha was the primary component of "gloss oil," a cheap media used for producing very-fast-drying interior enamels at the lower end of the market. Although better than raw rosin, limed rosin was friable, with poor water resistance, and so acidic that basic pigments tended to liver in the medium.[4] But a vastly improved product could be made by esterifying rosin with glycerol to produce ester gum.

Ester Gum and Resinates

Ester gum was made by combining rosin with about 10% glycerol. The resulting resin comprised abietic triglyceride with some mono- and diglycerides, together with some decomposition products and the unsaponifiable portion of the original rosin. Ester gum was first manufactured in the United States in 1893, although its use did not become widespread until some considerable time later (Langton 1943: 30). Ester gum was an improvement over limed rosin in terms of water resistance, outdoor durability, and livering. It was also used extensively in interior lacquer for items such as furniture, as it was compatible with nitrocellulose and most of the lacquer plasticizers and solvent mixtures. By the mid-1930s very pale gum was available, which could be used in whites and clear finishes. The superior color stability and alkali resistance of ester gum compared to unmodified rosin compensated for its relatively poor durability, and its low acidity rendered it suitable for use with basic pigments and for the manufacture of metallic paints (Chatfield 1953: 157). Ester gum was also used extensively in combination with tung oil to produce fast-drying paints and varnishes.

Other important products derived from rosin were metal soaps, or resinates, which were manufactured by reacting rosin with lead, manganese, and cobalt. Resinates were used as soluble driers and comprised the resin component of some coatings: ester gum and zinc resinate were the cheapest end of oil-based coatings. Resinates had a high metal content and high melting point and were readily soluble in drying oils and white spirit.

Another useful rosin derivative was maleic anhydride-reacted rosin neutralized with glycerol. As this tended to be marketed alongside manufacturers' phenolic resins, it is discussed in greater detail in chapter 5. Coatings based on these derivates were developed during the materials shortages of World War II, and they continued to be used for some years after, especially if durability was not a priority (Molloy and Riley 1948: 96–97).

Rosin derivatives formed the resin component in numerous oleoresinous finishes, and they were combined with phenol-formaldehyde and with alkyd resins and used in combination with tung oil. Increasingly sophisticated treatment of rosin meant that by the mid-1930s rosin derivatives had become important ingredients, and statistics published at that time suggested that over 20% of the total consumption of rosin in the United States went into paint and varnish manufacture, with only the paper and soap industries using more (Sanderson 1934: 711).

EXTERIOR PAINTS BASED ON TRADITIONAL MATERIALS

Exterior paints are divided into two categories: house paints that protect and decorate the main body of a wooden house, and trim paints that coat shutters, doors, and window frames. House paint tended to be white and trim paints colored so as to provide a contrast. Exterior house paints must protect surfaces that are exposed to extreme weathering and need to be flexible and durable. It is therefore essential that the raw ingredients are of the highest quality and that the paints are properly formulated.

Historically, exterior house paints consisted of nothing more than linseed oil and driers, pigmented with lead white paste and thinned with turpentine. These materials were bought by the decorator, who hand-mixed his paints prior to each job, altering the ratio of components depending on the end use of the paint; this practice continued until well into the 1950s (Singer 1954b: 58), long after the introduction of good-quality ready-mixed paints. Linseed oil in its raw, refined, and bodied states was also used by the manufacturer in the production of ready-mixed house paints, which became increasingly popular throughout the late nineteenth and early twentieth century. Ready-mixed paints were frequently pigmented with mixtures of lead white and zinc in order to obtain the best properties of both pigments. Zinc increased the weathering, durability, hardness, and opacity of the film, although if present in too great a quantity, it caused early cracking and flaking.

A house paint recipe dating from 1913 described a high-grade exterior paint comprising a paste made from refined linseed, lead white, American zinc, and magnesium silicate, which was thinned with turpentine, pale japan drier, and refined linseed oil. A cheaper grade was made by replacing some of the refined linseed for raw and some of the white lead with floated barytes and magnesium silicate. An even cheaper grade was made by using the second recipe and adding to every one hundred pounds of the mixture fifty pounds of whiting in oil, then thinning this mixture with four gallons of linseed oil, one and a half gallons of liquid drier, and one gallon of emulsion.

The emulsion helped to keep the pigment and extenders in suspension in the can. It was manufactured by making solutions of animal glue, borax, and sodium silicate, which were run into the paint. If the paint contained lead or zinc, no emulsifying assistant was reportedly required—probably due to the fact that basic pigments such as lead carbonate and zinc oxide could form soaps with the medium to aid suspension—otherwise rosin varnish could be added to aid emulsification. Emulsions had to be used very sparingly, if at all, in paints containing hygroscopic materials such as yellow ocher, raw umber, and raw sienna, or quantities of gypsum. Adding water to the paint binder increased the body, the effect of which increased even more if the pigment also had affinity for water (Troutman 1943: 368).

Historically, exterior paints were designed to chalk gradually in order to retain a clean, bright surface. However, chalking of lead films did not commence until some time after the coating had been applied, and paints pigmented with lead, or lead and zinc combinations, were susceptible to a serious defect: blackening of the film. The discoloration tended to be attributed to dirt retention or film failure (Vannoy 1943: 274) but was in fact most likely to be mildew or the reaction of pigments exposed to pollution. By the mid-1950s mold growth was a recognized problem, and formulators included greater quantities of zinc and fungus inhibitors such as phenyl mercury in their paints to prevent it (Singer 1954b: 68).

However, the problem of retaining a clean, bright surface was not really resolved until the introduction of anatase titanium dioxide in the mid-1920s. Titanium dioxide was used in combination with leaded zinc oxide and extender pigments. And although the resulting film ultimately failed by slow chalking and erosion, the chalking kept the coating clean and bright.[5] Further improvements were made in 1941 on the introduction of highly opaque rutile titanium dioxide pigments; the rutile form had much better hiding power and durability and performance superior to previous white pigments (Vannoy 1943: 274). The optimum properties were achieved by using a combination of rutile and anatase forms, which enabled the rate of chalking to be carefully controlled. By 1954 a standard formulation for high-grade exterior paints was a combination of lead, zinc, and titanium pigments with or without some hiding or extender pigments (Singer 1954b: 60). It should be noted that exterior paints are no longer designed to chalk.

THREE-COAT AND TWO-COAT PAINTING SYSTEMS

Historically, a three-coat system was used on new exterior woodwork: the primer to satisfy the absorption of the wooden substrate; the undercoat to provide a smooth, opaque surface that could be sanded; and the final coat to provide gloss and protection. Two-coat systems were recommended for repainting jobs only and were considered insufficient under any other circumstances. However, these accepted standards were challenged in the United States in the 1920s, when a boom in housing construction coincided with greatly increased labor costs. In an effort to reduce the cost of painting, speculators specified two-coat systems that omitted the undercoat, openly accepting that it was a temporary finish that would be repainted within a year or so: this trend caused much debate at the time. The Forest Products Laboratory, in Madison, Wisconsin, appears to have been the first to conduct rigorous research into the performance and use of two- and three-coat systems, the results of which were published in 1941 (Browne 1941). It concluded that the durability of a coating depended entirely on its thickness rather than the number of coats applied and that two coats were equal if not superior to three-coat systems if the appropriate materials were used.

However, paints intended for three-coat systems did not perform well if applied in just two coats, especially if it was the undercoat that was omitted. The priming layer was too translucent to hide the wood and also absorbed too much oil from the finishing layer. The final coat thus lacked opacity and was unevenly absorbed by the wood, resulting in a patchy, streaky appearance.

The lack of hiding power and the absorption problems caused by the omission of the undercoat prompted decorators to develop a different technique: they increased the pigment content of their paints, significantly reduced the spreading rate per gallon, and applied the paint in much thicker coats, which resulted in coats of sufficient thickness for good durability. A viscous priming layer was more satisfactory, as the accepted view that the priming layer should penetrate the wood was also being challenged at that time. Penetration was in fact found wasteful and even harmful, as the binder became separated from the pigment and existed alone in the cavities in the wood, while the pigment particles were left at the surface. Although in the early 1920s prepared paints were far too thin and had insufficient pigment content to be applied in this manner, this could be achieved with paste paints that the decorator mixed himself.

When carried out well, the results of a two-coat system were very durable and the savings in labor costs significant. As a result of this, two-coat work gradually became accepted, and paint manufacturers began to print two-coat instructions on their labels, initially on their paste paints and then on their prepared paints. However, a lack of understanding of the requirements of two-coat systems on the part of the manufacturer meant that decorators were instructed merely to omit the undercoat and to use the priming and final layers in the same manner as a three-coat system; no mention was made of reducing the spreading rates either, which resulted in many unsatisfactory jobs.

Manufacturers finally realized that in order to achieve a good two-coat system, viscous paints designed specifically to be applied in thick coats were required: satisfactory results could not be achieved by using paints intended for three-coat systems or by using special primer coats finished with traditional prepared paints. The increased thickness of these two-coat systems was achieved by using boiled oils in the binder and by increasing the pigment content (Browne 1941).

The increased use of bodied oils was prompted not only by an increase in labor costs but also by the scarcity of raw materials that occurred during both world wars. Before World War II exterior house paints were based on low pigment concentration ratios with a high percentage of bodied and raw linseed oil. The paint had many good properties: it was very durable and retained its clean surface, and the high percentage of raw oil made it easy to brush out. However, the raw oil content also made the paint ropy and stringy, and the combination of low pigment volume concentration and high oil content made the material costs high (Singer 1954b: 66). Therefore, in order to reduce the amount of oil required in paint, the United States developed an "emergency binder" ("Future of Wartime Paint" 1948). This binder had lower oil content and higher pigmentation. About 50% of the oil was heavy kettle-bodied linseed, which penetrated less and had better flow and leveling. Due to the high pigment and low oil content, the film had lower gloss. The durability was good, only slightly poorer than prewar formulas.

Wartime shortages encouraged manufacturers to look for less expensive alternatives to linseed oil, and although by the mid-1950s the standard binder was still linseed, it was being used in increasing amounts in combination with some kettled or bodied oils and driers. During the war a great deal of research was conducted into reducing the amount of oil needed in exterior paints and replacing linseed with alternatives. The postwar years thus saw the introduction of a great number of alternative oils, and soybean, oiticica, dehydrated castor, tall oil, fish, and safflower were used in greater quantities, often in combination with linseed.

TRIM PAINTS

Broadly speaking, trim paints are glossy, pigmented finishes that are used to protect and decorate exterior and interior wooden trim, including shutters, window frames, doors, and skirting boards. Although in many respects interior and exterior trim paints have similar properties, there are important differences. Ideally, all trim paints should possess easy brushing, dry quickly to a hard film, be self-leveling, have good color and color retention, and good gloss and gloss retention. But exterior paints must also be water resistant and must not fade on exposure to weathering (Singer 1954b: 74). Therefore, trim paints were formulated according to their end use, and they varied considerably in terms of cost, durability, and raw ingredients. It should also be noted that consumers in the United States accept a lower level of gloss than those in Britain, and the term *gloss* is thus a subjective one.

Historically, trim was painted with oil paint; if a glossy finish was required, a clear varnish could be applied. In the late nineteenth century the first ready-mixed enamels, enamel paints, and gloss paints were introduced. A vast range of materials could be used in their formulation, and they varied widely in terms of their quality and durability. Less durable coatings were marketed for interior use only, while others were suitable for both interior and exterior applications. The first glossy paints to be introduced were enamels. These were highly durable finishes that were used for exterior trim and interior trim and walls.

ENAMELS

Historically, enamels are associated with the highest-quality air-dry and baking finishes that produced an exceptionally smooth film with good build and excellent durability, especially where exposure is severe (Snelling 1953a). However, the term *enamel* was somewhat misused throughout the early part of the twentieth century in that it had a tendency to be applied to any paint with a gloss finish, whether that finish was based on oil, varnish, or synthetic resins.

Enamels were introduced in the late nineteenth century, and most comprised zinc oxide ground into a paste with varnish or oil, which was then thinned with a gum varnish or turpentine. For some authors, "genuine" enamels were those made of a pure oil binder, and they thus contended that any enamel containing resin or varnish should properly be described as an "enamel type" finish ("Terminology of Paint and Painting" 1941: 25). However, it is evident that the term was used to describe both pure oil binders and those containing varnishes and resins. Zinc oxide, which was introduced in 1850, was an essential pigment in the formulation of enamels. It could be ground incredibly fine, and on drying it rose to the surface of the film, producing a bright, white finish ("Enamels and Enamelling" 1923: 53).

Enamels were formulated for both interior and exterior use. Although the hardness and properties could be altered by changing the raw ingredients, the basic difference between them was that zinc oxide was ground into a soft dammar varnish for interior applications and a highly oxidized stand oil for exterior applications; the slow-drying nature of the stand oil binder was offset by its excellent durability.

A 1913 text describes a recipe for a traditional, moderately priced interior white enamel known as "china gloss" as French process zinc white ground into dammar varnish, with the addition of a small amount of lead white ground in bleached oil or dammar varnish. The paste was then thinned to a flowing consistency with dammar or very pale rubbing varnish. China gloss finishes were somewhat soft, and if interior enamels were to be rubbed and polished, the addition of a hard gum varnish was essential. The highest-grade interior enamels were made by grinding zinc in dammar to a paste consistency, beating the paste with turpentine, then slowly adding a very pale rubbing varnish (see plate 9).

An exceptionally high grade of exterior enamel might comprise condensed zinc white, highly oxidized oil, and turpentine, which were ground to a paste and then set aside in a covered container for at least two weeks. To this paste was added a kauri gum varnish, at the rate of forty pounds of paste to sixty pounds of varnish; the oil conferred elasticity and wear and the gum varnish high gloss (Uebele 1913).

In terms of application technique, enameling shared more with coach finishing than decorating. The enamel layer was the final coat, and a good job required careful preparation in the form of numerous undercoats. It was highly skilled, time-consuming, and thus expensive. One author writing in 1923 stated that ten undercoats were the least that should be applied, with fourteen or fifteen being ideal. Each coat was carefully rubbed down, and considerable drying time between each layer was necessary. The application of the final coat of enamel was not easy: the technique consisted of laying the enamel on the surface with as little brushing as possible. All enamels set so quickly that lap marks showed if the previous section was left for more than one minute. However, if too much enamel was loaded on the brush, sagging and lumps occurred. One author recommended heating the enamel if it was too cold to flow properly ("Enamels and Enamelling" 1923: 53).

A good enamel was self-leveling. The mirrorlike enamel followed perfectly the contours of the underlying layers: it was an unforgiving finish that accentuated flaws and imperfections. Although capable of producing an excellent finish, enamels had a number of disadvantages. The varnishes that early enamels were based on were thicker and heavier than linseed oil and were therefore not able to hold as much pigment, or the binder became too viscous to use; this, coupled with the use of the relatively translucent zinc oxide, meant that enamels lacked opacity, hence the need for numerous undercoats. The binder also lacked either durability or gloss, depending on the varnish used: short-oil varnishes were glossy but suffered cracking and ultimate failure, while long-oil varnishes, although more durable, had insufficient gloss. Manufacturers thus searched for a binder that had the durability of an oil paint and the gloss of a varnish.

It had long been known that linseed oil could be thickened to a varnishlike consistency by allowing it to stand for considerable lengths of time, or by heating in kettles, so manufacturers used the highest grade of varnish makers' oil and heated it in this manner. The resulting stand oil binder was almost colorless and had the consistency of strained honey. It was pigmented with zinc oxide, with a small amount of driers added, and by 1923 it was reportedly being used in the best grades of enamels ("Enamels and Enamelling" 1923: 53): these were the finishes classed as "genuine" enamels by many authors. Because the binder had greater body, it was possible to produce a good job with far fewer coats. It would appear that the use of stand oil as a binder was more widespread in the United Kingdom and Europe than in the United States. One author asserted in 1913 that grinding zinc oxide in oil rather than dammar was a tradition followed by Dutch, British, and German varnish makers (Uebele 1913).

However, even the improved enamels had significant drawbacks. The use of stand oil conferred a viscous consistency that was hard work to brush out, making the job slow and laborious. Furthermore, they still took some skill to apply properly. And finally, stand oil dried much more slowly than finishes containing varnishes and resins and could not be rubbed or polished for a week or two after application (*The Decorator,* July 23, 1923, 54). However, the use of enamel finishes increased dramatically during the first two decades of the twentieth century, as the home user recognized the advantage of a hard, lustrous finish that could be easily cleaned (Cayley 1926). In response, manufacturers introduced an increasing array of enamel-type paints that were easy to apply and covered in few coats.

Enamels and enamel-type paints were used for high-class interior and exterior trim and for interior wall finishes. They could be formulated into flat or glossy wall paints and were used where exceptional durability and washability were required, for example, in kitchens, bathrooms, and public buildings such as schools and hospitals. Flat enamels were made by replacing the refined linseed oil binder with a flat varnish ("Flat Paints" 1931: 663). Despite the fact that enamels produced an exceptionally fine surface that had extreme durability, their use had declined by the mid-1930s, mostly due to the increased availability of cheaper flat oil finishes (Symons 1933).

READY-MIXED GLOSSY PAINTS FOR INTERIOR AND EXTERIOR USE

The material costs and time involved in enameling meant that only the wealthy could afford it, and in reality, most exterior trim finishes comprised a raw or refined linseed oil binder to which bodied oils and high-grade oleoresinous varnishes were added. Although it was recognized that the addition of resin to a paint system contributed to its embrittlement and eventual breakdown, for exterior trim paints, the advantages of such an inclusion by far outweighed the disadvantages. By replacing a significant proportion of the raw linseed binder with bodied oils or varnishes, the pigments and extenders required to give the paint good body and durability could be reduced. This reduction resulted in better flow, gloss, and color retention. Exterior paints also needed to cover in one coat, which could be achieved only by increasing the body of the paint. Bodying was achieved via a combination of methods: although the varnish binder was already more viscous than raw oil, the body could be increased further by including water emulsions, heavily limed oils, or zinc stearate (Parker 1943).

The manufacture of exterior and interior ready-mixed gloss paints flourished throughout the early twentieth century, and by the 1920s numerous companies were marketing them. It was at that time that the confusion in nomenclature began in earnest. Companies advertised gloss paints, enamel paints, hard-gloss paints, high-gloss paints, oil-gloss paints, and varnish paints. Some were marketed for interior use; others were suitable for both interior and exterior applications. The differences between these types were not apparent, and it was all but impossible to make a distinction between them. Indeed, by 1947 one author had admitted defeat, stating that there was little point in trying to make the distinction between enamels and gloss paints, since although the formulations may at one time have been different, they must now be regarded as synonymous ("Treatment of Kitchens" 1947: 38). However, whereas exterior paints required considerable amounts of oil in the binder to obtain sufficient durability, interior paints could and indeed did draw on a diverse range of materials (see plate 10).

The quality of interior glossy paints varied considerably. Durability was not essential, so the cheapest grades might consist of nothing more than pigment ground into a cheap spirit varnish based on rosin, dissolved in benzene or coal tar naphtha (Coffignier 1922). Although coatings based on spirit-soluble varnishes were touch dry in around thirty minutes and had brilliant initial gloss, the gloss deteriorated so rapidly that it could really only be considered temporary. Other low-grade materials added to trade varnishes included shellac, celluloid, and gun cotton (Matthieson 1906).

The use of oil varnishes increased the durability somewhat, although the quality of course depended on the quality of the constituent ingredients. A typical inexpensive gloss paint might comprise a ready-mixed oil paint of indifferent quality to which some outside varnish had been added: although such finishes initially looked good, they too deteriorated rapidly (*The Decorator,* August 22, 1925, 135). There were, of course, plenty of intermediate-quality binders used in the manufacture of interior gloss paints. Limed rosin, ester gum, and maleic anhydride-reacted rosin were all used extensively in the formulation of interior gloss paints. Tung oil combined with ester gum, sometimes in admixture with linseed, was a useful binder for reasonably durable glossy paints. Indeed, ester gum was an essential component in the formulation of glossy paints, and it remained so until well into the 1950s, at which time it was superseded by far superior alkyd resins.

INTERIOR FLAT OIL-BASED WALL PAINTS

Ready-mixed flat paints became widely available in the 1930s (Adair 1938; see plate 11). They represented a considerable advancement in the development of durable, washable, interior wall paints. A flat finish was achieved in one of two ways: by reducing the binder quantity and increasing the pigment content, or by the addition of a flatting agent. Flat oil paints were in direct competition with oil-in-water emulsions, distempers, and casein paints but were chosen where greater durability was required. Their use declined dramatically in the late 1940s and early 1950s, when emulsion paints based on synthetic polymers became popular, but they were still available until well into the 1950s.

Prior to the introduction of ready-mixed flat paints, producing a flat finish from pigment paste, thinners, oils, and driers was no easy task and could be achieved only by flatting. A flatting coat consisted of a lead or lead and zinc pigment paste heavily thinned with turpentine to a brushing consistency: no other oil or binding media was included that might impart gloss to the film. The resulting pigment content was very high, around 90% by weight ("Flat Paints" 1931: 663). The flatting was applied to a very oily undercoat that had been applied the day before and was still tacky: the excess oil provided a binder for the flat coat and helped the pigment adhere to the surface. Once the turps had evaporated from the flatting layer, dry pigment was left on the surface ("Flat Oil Finishes" 1957: 43). Flatting was a very skilled task. The coating was finished by stippling but set so quickly that at least two men were required: one to apply the flatting and the other to stipple. It was difficult to apply over large areas and almost impossible to apply if there was a draft, as the finish set so quickly that is was virtually impossible to brush out or stipple. It did not work well on porous surfaces, and flashing (glossy patches) was a common problem. Moreover, the action of oxygen on the tiny amount of binder present rendered it brittle, weak, and susceptible to failure: it was unsuitable for exterior applications for these reasons (Harrison 1931). The underbound coating also had poor washability and absorbed dirt easily.

Achieving a flat finish became easier with the increasing sophistication of paints prepared by the manufacturer. The first ready-mixed flat wall paints were introduced in the mid- to late nineteenth century. These comprised around one-third lead white and two-thirds zinc ground in refined linseed oil. A small amount of pale drier and varnish was added, and the paste was thinned with turpentine. A flat finish could be achieved by adding a small amount of borax solution (Uebele 1913). By the 1930s many companies were producing good-quality flat wall paints or flat oil finishes. Better pigmentation, the development of binders that themselves dried to a matte finish, and the use of heavier diluents and special retarding agents all contributed to the introduction of good-quality flat paints with few of the drawbacks of traditional flatting (Harrison 1931).

Improved pigmentation was effected by the introduction of lithopone, which came into general use in paint manufacture in the 1880s. Lithopone had excellent covering power and was widely used to pigment both water- and oil-based flat wall paints ("Flat Paints" 1931: 663). Lithopone contains varying proportions of zinc sulfide and barium sulfate, with a small amount of zinc oxide. It was used for interior applications only, as it turned gray on exposure to sunlight. Lithopone was also an excellent pigment for tung oil, a much-improved binder for flat wall paints. Tung dried very quickly to a tough, durable film that could be scrubbed within forty-eight hours (Jennings 1925), and the heavy petroleum thinners used to reduce it also contributed to its pleasing, very matte finish (Uebele 1913).

Another method of imparting a flat finish was through the use of flatting oils. These oils were made by cooking an oil or mixture of oils until they had enough body that they could be thinned with considerable quantities of solvents without becoming excessively fluid. The creation of false body enabled the coating to be brushed out exceptionally smoothly. Such a binder reportedly offered greater opacity than the amount of pigment might suggest. Flatting oils were more commonly used in the United States than in the United Kingdom ("Flatting Oils" 1932: 6).

Ready-mixed flat paints were a huge improvement for the professional painter: they produced a soft, velvety finish and were reasonably durable and washable. From the user's perspective, they were the most economical, as they were applied with a large five-inch brush, had good opacity and flow, and did not require being laid off with a brush or stippled, so labor costs were reduced (Adair 1938). Although more durable than either water paints or flatting, ready-mixed flat wall paints were not suited to surfaces that required regular washing or were subject to atmospheric moisture. For these applications, the gloss and enamel finishes described above were employed.

NATURAL RUBBER IN HOUSEHOLD PAINTS

Rubber latex is the naturally occurring elastomer, polyisoprene, which derives from many flowering plants: the most commercially important source is the Para rubber tree *(Hevea brasiliensis)*. The latex is harvested by making small slashes in the trunk and collecting the material that exudes from the wound. Rubber latex is a milky white liquid that comprises complex organic compounds such as proteins, alkaloids, starches, sugars, oils, tannins, resins, and gums, suspended in water. The latex is processed before formulation into rubber products: the processing technique chosen depends on end use. Isoprene is also manufactured synthetically, as are many other monomers that are used to make a range of synthetic rubbers.

Natural rubber is an incredibly diverse material that, depending on the chemicals it is compounded with, can be formulated into a range of materials. The rubber is purified, then chemically cross-linked by the vulcanization process to become the highly extensible material that we are familiar with: from car tires and gaskets to bowling balls, surgical gloves, and elastic bands. Rubber has also been formulated into coatings, the most well known of which are probably the chlorinated rubbers used for industrial applications where exceptional durability and chemical and water resistance are required.[6] However, a less-well-known application for rubber was as an ingredient in oil- and water-based household paints, especially during the 1930s and 1940s, and possibly the 1950s. However, it is important to note that "rubberized" paint does not necessarily refer to a paint that contains natural rubber and is in fact more likely to be describing a paint based on a synthetic polymer emulsion (Wolf 1953).[7] This confusion in terminology arose because styrene-butadiene, the first synthetic polymer to be used in the formulation of household emulsion paints in the United States, was originally developed as a rubber substitute. This also accounts for the fact that styrene-butadiene and indeed all subsequent household paints based on synthetic polymer emulsions in the United States were and still are called "latex" paints.

Rubber was incorporated into paints by a number of methods. Solutions of solid rubber in white spirit were used as a replacement for natural resins in oil-based paints; rubber solution could also be emulsified with a protein such as glue or casein and added to water-based distempers; and finally, a dispersion of latex in water could be added to distempers ("The Uses of Rubber" 1934). Natural latex was also used as a plasticizer for casein paints (Warth and Browne 1939: 338).

Harvesting Rubber Latex

Raw latex is collected from the tree in small cups. It immediately begins to coagulate (due to evaporation and oxidation) and is susceptible to bacterial attack, so a preservative and stabilizing agent is added. Historically, ammonia was used for this purpose, although today ammonia is supplemented with secondary preservatives. The ammonia is added at concentrations of around 0.7% to 1.0% by weight and not only prevents bacterial growth but also helps to maintain colloidal stability (De and White 2001: 19). The latex is then sieved to remove foreign matter such as bark and leaves, blended to ensure a consistent product, and stored in collection tanks. The latex can then be processed into solid rubber or latex concentrate, both of which have been used in household paints.

Manufacture of Solid Rubber

To produce solid rubber, the latex is processed into "ribbed smoked sheets" and "crepes." The solutions of solid rubber used in household paints would have derived from high-quality pale crepe-rubbers, as ribbed smoked sheets have poor color. The best-quality crepes are manufactured by removing the yellow carotenoid pigments from the stabilized latex. Removal is effected by bleaching and/or fractional coagulation. The first fraction is yellow in color and unstable: this is removed and the remaining latex processed further. The processed latex is then coagulated and pressed into slabs. The slabs are passed through crepe rollers while being thoroughly washed and are left to dry in hot air for about two weeks. The pale crepe is then graded visually and pressed into bales for storage and transportation (De and White 2001: 22–23).

If the crepe is to be made into a rubber solution, it is masticated to break down the large polymer units, which allows the rubber to be solubilized without gelling. For use in household paints, the crepe was reportedly milled with a drying medium (usually cobalt linoleate), then mixed with white spirit ("Rubber Paints" 1938: 68). The resulting rubber solution was compatible with the usual drying oils used in paint and varnish manufacture but incompatible with resins or fossil gums. The rubber solution was therefore used in place of the gum or resin component and not in combination with it.

Manufacture of Paints from Solid Rubber

It is evident that rubber solution was available in different forms, depending on end use. For example, a 1954 booklet produced by the British Rubber Development Board described oxidized, cyclized (or isomerized), and chlorinated rubber for use in paints. Although chlorinated rubbers are restricted to industrial coatings and therefore are not discussed here, both oxidized and cyclized rubbers were recommended for use in household paints: the former in flats and undercoats, the latter in glossy, hard finishes resistant to water, chemicals, and condensation. Both oxidized and cyclized rubber were dissolved in white spirit to make paint vehicles, but the hard nature of cyclized rubber meant that it usually required additional plasticizer. Moreover, although oxidized rubber dried quickly to a flat finish, cyclized rubber dried by evaporation, followed by a slow oxidation process that took several weeks. The coating was not fully resistant until this process was complete, which made recoating difficult, as the first layer could be "plucked" from the substrate. In order to counteract this, around 20% of a drying material such as alkyd was included ("Rubber Paints" 1954: 44). The resulting coating was hard, with high gloss, low moisture permeability, and excellent steam resistance, making it suitable for kitchen and bathroom paint and other interior decoration applications.

However, it is evident that rubber solution found its greatest use in the formulation of undercoats and flat paints. When used in equal proportions with oil, it made a paint with good flow and leveling, which produced an exceptionally smooth, flat finish. Undercoats based on it did not need sanding, and top coats did not require time-consuming stippling to produce an even finish ("Rubber Paints" 1938: 68).

Rubber was also combined with other compounds to produce coatings. For example, chemists had reportedly been searching for a binding medium with the hardness of celluloid and the flexibility of rubber for some years and by the mid- to late 1920s had patented coatings based on colloidal solutions of rubber and cellulose. By the early 1930s these rubber-cellulose coatings were being advertised in the British trade press. For example, Rucel and Plax, finishes of this type, were marketed as waterproof varnishes that could also be used as a medium for paints, washable distempers, lacquers, stains, preservatives, and printing inks (Morris 1931).

Rubber deteriorates when exposed to UV light, and in order to mitigate this fault, is it likely that inhibitors were added. However, there is just one reference in the trade literature to the inhibition of rubber used in paints. A DuPont/ ICI patent from 1930 describes a process for making paint comprising solution rubber and oil, with or without drier, thinned with a volatile solvent to a suitable consistency, with the addition of an inhibitor such as 1% to 10% quinol to retard the catalytic depolymerization of the rubber ("Latex Finishes" 1930: 76).

In addition to being used in solvent-borne paints, rubber solution was incorporated into water-based distempers. To render the solution compatible with water, it was emulsified with glue or casein before being added to the paint. Dispersing agents, including stearic acid, colloidal clays, glues, and soaps, or admixtures of these and other substances, were required ("Dispersed Rubber" 1931: 29). Rubber was also incorporated into water-based paints in the form of a latex concentrate, for which the rubber underwent a different processing procedure.

Manufacture of Paint from Latex Concentrate

Although it is possible to use sieved, stabilized raw latex directly for many applications, including as an addition to distemper, it is usually concentrated in order to increase the dry rubber concentration, reduce tanking and transportation costs, and, depending on the technique used, remove nonrubber components. The latex is concentrated by centrifugation (most common), creaming, or evaporation. While newly tapped rubber latex has a dry rubber content between 30% and 40%, this rises to around 60% in latex concentrate (Morton 1987: 184). Latex concentrate was added to distempers to improve washability and water resistance and was also used as a plasticizing resin for casein paints in the 1930s.

History of Rubber in Paints

It is evident that aqueous dispersion of rubber was considered as a paint ingredient some time before solvent-based solutions were. The first patent describing a method of dispersing rubber in water was taken out in England in 1836: a water dispersion of rubber was obtained by immersing the rubber in a solution of ammonia and simply waiting for the ammonia to evaporate, leaving the rubber dispersion behind ("Dispersed Rubber" 1931). These natural latexes were assessed for their suitability in emulsion paints, and as early as 1924 paints based on natural rubber latex, with casein as the thickening and stabilizing solution, were patented (Fowler and Westerhoff 1952). However, these early attempts to use aqueous dispersions of rubber latex were not commercially successful, largely because rubber latex was a new material with very complicated colloidal properties, and the correct compounding for desirable stability, viscosity, and other properties was not fully understood. Furthermore, the polymer itself did not have desirable drying, aging, and pigment-binding properties and was especially susceptible to degradation by UV light.

By the early 1930s coatings based on rubber became widely available, and they are discussed with reasonable regularity throughout the decade. The most important patents reportedly originated from the United States ("Dispersed Rubber" 1931). By the early 1950s their use in household paints was largely superseded by synthetic alternatives, in both oil- and water-based paints.

Properties of Paints Containing Rubber

Industrial finishes such as chlorinated rubber are most renowned for their exceptional water and chemical resistance. Chlorinated rubbers produce flexible, damp-proof, durable films that are impermeable to moisture and are used in applications such as swimming pools, marine paints, and protecting metals exposed to harsh chemicals.

The inclusion of rubber imparted similarly good properties to household paints, and it was used to most advantage in the formulation of oil-based flat wall paints and undercoats. Oil paints incorporating rubber solution had good flow and leveling and produced an exceptionally smooth surface that was completely free from brush marks. Coatings containing it were also highly washable. The inclusion of rubber prevented heavy pigments from hard settling in the can—a significant problem with oil-based paints that was not easily reversed. However, the presence of rubber enabled any settled pigment to simply be shaken or stirred back into the paint.

The paints dried quickly enough to prevent dust deposition but had sufficient open time to allow large surface areas to be covered without lap marks ("Rubber Paints" 1938: 68). Moreover, they dried to a good matte finish, with no visible brush marks, eliminating the need for time-consuming stippling or flatting. Although used primarily in undercoats and flats, rubber could also be added to ready-mixed oil paints (to the order of 15% to 20%), where improved gloss, flow, and leveling were claimed ("Rubber in Paints" 1942: 29). Rubber was also used as an inexpensive ingredient in gloss and enamel paints: because the good flow exhibited by traditional enamels was achieved through the use of expensive fossil gums and resins, it made sound business sense to experiment with the considerably cheaper rubber ("The Uses of Rubber" 1934). The cyclized rubber described by the British Rubber Development Board produced hard, glossy films with excellent steam and condensation resistance, and a photograph in the leaflet depicted a bathroom painted with it ("Rubber Paints" 1954: 44).

Rubber also imparted durability, water resistance, and scrubbability to water-based paints, and it reportedly allowed the manufacture of distempers that were truly washable (Morris 1931). The paint had slightly different properties depending on whether the rubber was added as an emulsified solution or a latex concentrate: the latter produced harder, more durable films, but the distemper is not so easy to handle, and the brush clogged rapidly unless it was washed immediately after use ("The Uses of Rubber" 1934).

Availability and Use of Rubber Paints

The extent to which rubber was used in household paint formulation is by no means clear. The first articles discussing rubber in household paint appear in the British trade press in the early 1930s, and rubber is mentioned intermittently throughout the decade. A 1938 article in *The Decorator* claimed that many leading manufacturers supplied both gloss and flat finishes that incorporated a proportion of rubber ("Rubber Paints" 1938: 68). Other authors, however, claimed that it was not much used due to its tendency to degrade on exposure to UV radiation (Gardner 1935).

However, decorative coatings containing rubber were certainly advertised. For example, the Sigmarine company's Tornol Rubber Paint, available in 1933, was marketed as a high-gloss, very resistant paint intended for ordinary interior and exterior decorating work, with other grades for auto finishes and industrial uses (*The Decorator,* November 15, 1933, 54). The author of a 1948 article contended that before the outbreak of World War II many manufacturers included rubber in their flat paints and undercoats, but the practice ceased during the war due to material shortages. But by 1948 rubber was once again in good supply, and as the availability of oils, gums, and resins was still restricted, it is probable that rubber was included once more in household paints during this time (Stokes 1948). However, by the late 1940s and early 1950s the use of rubber declined: the flat wall paint market began to be dominated by emulsion paints based on vinyl polymers, and oil-based decorative paints made increasing use of oil-modified alkyds.

Chapter 3

Traditional Water Paints

Historically, water paints were used to protect and decorate interior and exterior plaster, masonry, and wood, although in more recent years waterborne glossy paints for wooden trim have also been developed. The earliest and simplest water paints were whitewashes and distempers, which have been around for centuries. However, these early paints lacked durability and were frequently fortified with waxes, proteins, fats, and oils in order to increase washability, water resistance, and weathering. This practice was continued by the manufacturers of the first "washable distempers" in the late nineteenth century, who formulated coatings with varnishes, oils, and casein to improve durability. The use of casein increased dramatically throughout the first half of the twentieth century: it was used as both the sole binder for interior and exterior wall paints and an emulsifying agent for the oil emulsions that were introduced in the late 1930s.

Water paints produce attractive soft, flat finishes, which are moisture permeable. They are inexpensive, easy to apply, and easy to clean up. Although today they have largely been displaced by modern paint systems based on synthetic resins, they are still used where their unique properties are required, in particular, for the conservation of historic buildings.

LIMEWASHES

Limewashes are the earliest and simplest water paints. They were used extensively to coat interior and exterior walls of homes, barns, and outbuildings. A great advantage of limewash is its moisture permeability, which prevents the buildup of condensation and associated problems.

Limewash was the cheapest coating available and was easy to make: limestone or chalk (calcium carbonate) was burned in a kiln to produce quicklime (calcium oxide). The quicklime was then slaked by the addition of water (calcium hydroxide) and the slaked lime further diluted with water; some pigment could be added if a tint was required, or chalk could be added to make whitewash. The coating was applied with a broad brush and was initially transparent. This made it difficult for inexperienced users to judge how much to apply, and a common problem was to apply far too much. The coating dries by absorbing carbon dioxide from the atmosphere, which reverts the calcium hydroxide to carbonate: the coating gains opacity over the few days it takes this process to complete.

The above recipe describes limewash in its simplest form, but without modification the resulting coating is soft, with poor water and abrasion resistance. It was therefore common to add wax, fat, or protein in order to increase the durability and water resistance of the film (Warth and Browne 1939: 315). One of the most common additions was the animal fat tallow. The stearic acid in the tallow reacts with limewash to form calcium stearate. By increasing the amount of calcium stearate emulsified in the limewash, better adhesion, the sealing of excessive porosity in the substrate, and improved water resistance are observed. Modification with tallow extends the life of limewash from about one year for unmodified types ("Limewash Coatings" 1947: 36) to three years ("Water Paint" 1946: 59). Glue, starch, and flour paste were also suggested as modifiers, along with formaldehyde, which acted as a preservative.

Due to impurities, limewash has a yellowish tint, and it was common to add some blue pigment to counteract this. The cheapest pigment was synthetic ultramarine, which was ground in water and added to the mix; most manufacturers produced a range of "lime" pigments to tint limewash to its characteristic pastel shades ("Limewash Coatings" 1956: 53). However, pigmentation was the cause of many problems, including patchy drying, caused by insufficient mixing, and uneven absorption of the underlying layer.

The major drawback of limewash was the matter of renewing the coating: it did not provide a good ground for any other type of finish and was costly and difficult to remove. In most instances once limewash had been employed, there was no opportunity for using anything else, as the friable limewash surface simply flaked away, taking the new coating with it; steam was often the only effective method of removal. Moreover, the highly alkaline nature of the finish precluded the use of a saponifiable coating. Limewash could only really be coated with more limewash, with the result that coats of tremendous thickness could be built up over the years.

Limewashes played an essential role for interior and exterior decoration in both the United States and the United Kingdom and were far more widely used than oil, which was considered a luxury product. Limewash remained one of the most popular paints until the end of the nineteenth century, and it was still being widely used in the 1930s. By the 1950s its use had been largely superseded by more modern coatings, although it was and still is used today for certain specialist applications.

DISTEMPER

The word *distemper* has been used to describe a range of water paints, but true distemper consisted of nothing more than calcium carbonate (chalk or whiting) in animal glue or dextrin. It was sold in the form of a dry powder that was mixed with hot or cold water just prior to use. After limewash, it was the cheapest coating available. The popularity of distemper arose from its inexpensiveness; the ease with which it could be applied; its low odor; its quick-drying soft, flat finish; and the fact that the substrate required no special preparation (Warth and Browne 1939: 325–26). All these factors ensured that distemper was used extensively on interior walls and ceilings throughout the nineteenth century and well into the twentieth (see plate 12). Distemper had other advantages over oil-based paints in that it was resistant to hydrolysis and was moisture permeable. This made it useful for coating new and damp plaster—conditions that will destroy an oil paint—and it thus found extensive use as a temporary coating for the walls of newly built homes that had not yet dried out.

The main disadvantage of distemper was the softness of the coating and the fact that it remained soluble in water. The primary ingredient was chalk, which was simply dispersed in animal glue or a similar medium; minimal glue-binding material was present, and as the binding medium did not oxidize, it remained water soluble for its life. Furthermore, the proportions of glue to whiting had to be correct: too much glue, and the coating cracked and flaked; too little, and it became friable and deteriorated rapidly. Distemper's solubility in water made it virtually impossible to recoat, as subsequent coats simply swelled and pulled off the underlying layer. Because it remained water soluble, it was easy to remove (although removal was labor-intensive), and it needed to be thoroughly washed away with water before repainting could proceed. As distemper aged, delamination from the substrate also occurred, with resulting peeling and flaking.

As was the case with limewash, distemper was usually fortified with proteins, waxes, oils, or resins to improve its durability. Historically, this was done by the user, but the first factory-made fortified distempers were introduced in about 1908 ("Distempers" 1938: 51). The simplest method of making distemper washable was by replacing the animal glue binder with casein and lime (Warth and Browne 1939: 326), thus making what in reality was a casein paint. Indeed, the confusion surrounding the nomenclature of waterborne paints becomes apparent at this time, a problem exacerbated by the tendency of the British industry in particular to call any fortified waterborne coating "washable distemper." In its July 15, 1933, issue *The Decorator* listed three main types of distemper: whiting and powdered glue supplied in dry form; whiting and, in some cases, lithopone with glue or casein as the binder, supplied as heavy paste; and emulsion type (washable distemper), comprising whiting and lithopone with linseed oil emulsion as a binder ("Composition of Distempers" 1933: 67). These might more accurately be described as true distemper, casein paint, and oil-in-water emulsion paint. The confusion that this blanketing of terminology could cause was considerable: following an article that appeared in a 1938 issue of *The Decorator,* in which distemper was (quite correctly) described as lacking durability, Sisson's Brothers felt compelled to reply, stating that its Hall's Distemper was in fact an oil-bound water paint, made with the finest ingredients (*The Decorator,* October 15, 1938, 59).

Despite the availability of more durable alternatives, distempers were still being widely used in the late 1930s—in both powder and paste form—as they were eminently suitable for low-cost coatings on new plaster walls ("Distempers" 1938: 51). From the late 1920s, washable distempers began slowly to be replaced by flat enamels and flat oil paints for use on walls (Cayley 1927). And these began to be replaced by oil-in-water emulsions in the late 1930s and early 1940s.

CASEIN PAINTS

Casein is a versatile paint ingredient that has been used to reinforce traditional water paints, as a binding medium in its own right, and as a stabilizer for a range of resins in water-based systems: from glycerol phthalate and natural rubber latex to styrene-butadiene. Casein is a phosphoprotein of milk. It is precipitated when unpasteurized skimmed milk is treated with dilute acids such as hydrochloric, lactic, and sulfuric. It is insoluble in water and solvents but dissolves readily in alkaline conditions. However, highly alkaline or acidic conditions cause the protein to decompose, so the pH of the paint system containing it needed to be carefully controlled. Casein varied considerably in quality, the primary influencing factor being the degree of care and chemical control taken during manufacture: tight controls commanded premium prices (Radcliffe 1943: 463–73). Casein is a versatile ingredient that has a long history in paint production. Indeed, ancient Hebrew texts make reference to milk protein as a paint binder, and it had long been known that the addition of milk to whitewash and distemper improved the ease of application and durability of the paint (Radcliffe 1943: 461). The first patents for casein in paints appear in the mid-nineteenth century (Scholz 1953), and the first factory-made casein paints became available in the United States in the 1890s, around the same time as other ready-mixed paints based on oils (Warth and Browne 1939: 315). However, casein paints did not really gain importance until they became available in paste form around 1930.

Early casein paints were sold in powder form; the powder was soaked in water to swell the components, then thinned with more water prior to use. They were pigmented with what would be termed extenders in other paint systems: namely, whiting, talc, chalk, magnesium silicate, blanc fixe, and china clay. Mica was often added to improve the brushing properties (Radcliffe 1943: 468). They were thus transparent when wet but formed an opaque film when dry. By the late 1920s, however, more opaque pigments started to be added, including zinc oxide, lithopone, zinc sulfide, titanium barium pigment, and titanium dioxide. The resulting finishes had greater opacity and were more akin to oil paints in terms of appearance (Scholz 1953). These improved finishes, which appeared in 1928, were available in either powder or paste form; both types are discussed in detail below.

Paste paints reportedly took two quite distinct strands in geographic terms: whereas in the United States, paints were predominantly casein bound, supplemented with additional binding agents, Europeans tended to use casein as an emulsifying agent for oil-in-water emulsions (Warth and Browne 1939: 316). However, there was often no clear distinction between the two types, as it was simply a matter of altering the ratio of ingredients: in order to obtain an emulsion, the ratio of oil to casein must be greater than two to one, which ensured that there was enough oil present to produce a coalescence of the oil phase on drying (Scholz 1953). If the ratio was less, the paint was considered casein supplemented with additional binding media.

The use of casein in paints was undoubtedly driven by the shortage and expense of vegetable oils. Britain imported most of its vegetable oils, and the severe shortages following World War I, coupled with national trade barriers, encouraged manufacturers to use casein. And for the United States in the grip of the 1930s depression, casein proved popular due to its low cost per unit of covering power (Radcliffe 1943: 462). Moreover, most efficient use of the available vegetable oil could be made if it was emulsified in water rather than used as the sole binder.

Manufacture of Powder and Paste Casein Paints

As long as casein paints were still essentially washable distempers, they were always supplied in powder form. However, with the introduction of improved casein paints containing opaque white pigments, their development split along two lines: powder and paste. Powder paints comprised dried casein, lime, and mineral pigments; due to the presence of lime, all the pigments used had to be alkali resistant. On the addition of water, the lime solubilized, thus providing the alkaline conditions required to solubilize the casein. Dry powder paints would keep for a long time, but once mixed with water, they had to be used the same day or microbial and bacterial attack would occur, with unpleasant results. Indeed, in the United States, federal specification required that such paints must be able to stand for eighteen hours after mixing with water without emitting an offensive odor (Radcliffe 1943: 469). Coatings derived from powder paints form an insoluble film on drying, due to excess lime in the paint film.

When in paste form, the casein was already in solution, which brought about a range of improvements in terms of ease of use but also a host of new problems that had to be overcome. The casein solution had to remain stable over a period of time, meaning that the alkali it was dissolved in must not cause precipitation if present in excess; it had to contain preservatives to prevent mold and bacterial attack; it had to maintain the pigment in suspension over time; and it had to produce a pleasing, durable film that was easy to apply. This was a lot to ask of one paint binder, and it is at this point in paint manufacturing history that the list of additives required becomes extensive. Paste paints were made by dissolving casein in hydroxides or salts of weak acids of alkali metals or ammonia. Indeed, one of the earliest patents included ammonia as the solvent (Radcliffe 1943: 469). This meant that the paint itself was caustic, which limited the type of packaging that could be used, restricted the pigment range, and also posed health and safety risks for the decorator.

There was evidently much discussion about the relative merits of powder and casein paints during the 1930s. Because powder paints were inextricably associated with low-cost distempers and whitewashes, they gained a reputation for being lower grade than paste paints. This problem was exacerbated by the fact that they were cheaper to buy than paste paints: not because they comprised lower-quality materials but because, being in dry form, they required smaller containers and took up less storage space. Powder paints' reputation for being of lower quality was certainly undeserved, as there was reportedly little to choose between them in terms of durability: in fact, powder paints were rather more moisture resistant than paste, due to the presence of excess lime, which acted as a hardener (Scholz 1953). In 1936 one manufacturer of powder paints criticized paste manufacturers in an advertisement that pointed out to consumers that they were simply paying paste paint manufacturers for water: "water $24.00 per barrel—reasonable on a desert—outrageous on a paint."[1]

The addition of an emulsified drying oil, varnish, or resin to casein paint overcame many of its weaknesses, improving the durability and water resistance. This led to the development of "oleocasein" paints, which were intended to be as durable as equivalent solvent-borne paint systems. However, as discussed above, there was no clear distinction between oleocasein paints and emulsion paints that used casein as an emulsifying agent, and it was thus virtually impossible to ascertain whether a certain paint was an oleocasein paint or an oil-in-water emulsion stabilized with casein (Radcliffe 1943).

Additives to Powder and Paste Casein Paints

Once water paints became available in paste form, numerous additives were required to ensure stability, ease of handling and application, and preservation. It was also necessary to increase the viscosity of the paint, which could be achieved by replacing part of the casein with dextrin, soluble starch, algins, alum, or wax soap. Although these additions made the paint easier to apply, it was at the expense of its water and washing resistance, as many remained water soluble in the dried film. Plasticizers were also added to prevent the coating from peeling and cracking; in the 1930s the most common plasticizers were dextrin, glycerol, glucose, sorbitol, castor oil, sulfonated castor oil, rubber latex, and rubber derivates. Again, the first four of these are water soluble and thus affected the ability of the film to be washed.

Foaming is a problem with all waterborne paints, and various ingredients were incorporated into casein paints to overcome this. Although effective as an antifoam agent, pine oil had an unpleasant odor and was also an irritant. Suggested alternatives included compounds such as diethyl phthalate and octyl alcohol (Warth and Browne 1939: 338). Preservatives were a necessity in paste paints, and they were sometimes added to powder paints also. Finding an appropriate preservative was also a major problem; a text from the early 1940s lists a huge number, including borax, tribromophenol, sodium trichlorophenate, parachlorometacresol, sodium and ammonium fluorides, theymol, chloroisothymol, sodium salicylate, salicylic acid, sodium orthophenylphenate, sodium 2-4-5 trichlorophenate, and phenol (Radcliffe 1943: 471). Formaldehyde was often added to increase the hardness and durability of the film.

Indeed, washing a freshly painted casein paint film with formaldehyde was a known method of increasing its washability and hardness.

Pigmentation of Powder and Paste Casein Paints

In technological terms most of the pigments commonly used in oil paints could be used in dry and paste casein paints, but in reality the selection was made from a smaller group. There was no technical advantage to using chemically active, basic pigments such as white lead and zinc oxide: both were not only expensive but heavy as well, and tended to settle in the can. Furthermore, lead in particular is difficult to disperse in water. The most important white pigments were thus zinc sulfide and titanium whites (Warth and Browne 1939: 329). Any colored pigments incorporated into powder paints had to be alkali resistant both in dry form and in solution, as an excess of alkali was present; this did limit the pigmentation options somewhat. Suitable pigments included earth pigments, cadmium reds and yellow, iron oxides, and the like. A poorly dispersed pigment resulted in an unsatisfactory film, manifest as bright spots of aggregated pigment. In response to this, manufacturers produced pigments that were easily dispersible in water. The advent of paste paints increased the range of pigments available for use in casein paints, as paste paints did not contain lime, so there was no excess alkali. As a result, the consumer was offered a range of brightly pigmented paints that had not been available in powdered form (Radcliffe 1943: 468–72).

Properties of Casein Paints

Casein paints have a number of good properties that ensured their popularity and widespread use. They were relatively inexpensive, had excellent opacity, could be applied to a range of surfaces without special preparation, and were easy to apply; in addition, they dried quickly, to a flat finish with clean, bright colors, and were easy to clean up. Furthermore, the paints were moisture permeable and therefore useful where passage of moisture was desirable, such as a coating for damp plaster. Once dried, casein paints form a hard adhesive film, so they were a suitable base for the subsequent application of oil or other types of decorative finishes.

Although casein paints were more durable than whitewash or distemper, both of which rubbed off if abraded, they were less durable than the oil paints they were in direct competition with and did not fare well if exposed to damp or wet conditions. Although casein paints were inferior in terms of durability to oil-based flat wall paints, they possessed unique properties that made them suitable for certain decorative uses. Casein paints offered bright, clean shades that simply were not attainable with the alternative oil-based paints. Oil paints discolor, a fault especially pronounced in interior applications where the film is not exposed to light. To counteract this effect, white oil paints were often tinted with blue, which introduced a grayish cast and resulted in muddy, muted colors; the colors became even murkier as the paint aged. Casein, however, is practically colorless once dried and remains so on aging. Furthermore, due to its excellent binding properties, very little casein is required in the coating; this means that there is only a small amount of binder present to absorb light. The film thus reflects a high proportion of the light that falls on it: indeed, the reflectance value of the film is very close to the pigment itself, which accounts for the beauty and clarity of the finish (Atwood 1934: 48).

Casein paints also dried to a very hard film that did not collect dust in the same manner as softer oil-based paints. So although they could not be washed to the same degree as oil paints, they did not collect as much dirt in the first place. Furthermore, casein produces a very flat film, which virtually eliminates defects in walls. This property was enhanced by the indirect lighting that became popular in homes in the 1930s. And it was for its very flat properties that casein paint reinforced with tung and other oils was chosen to decorate the exterior of the buildings at the Chicago World's Fair in 1933 and 1934 and the New York World's Fair in 1939 (Radcliffe 1943: 462); the buildings were illuminated at night, and a very flat finish was thus essential to prevent unsightly reflections. Casein paints achieved a real boost in the mid-1930s as a direct result of their use in the Chicago World's Fair.

Although the cost of casein was higher than that of equivalent binders at the time, making casein paints more expensive to manufacture, their good covering power meant that it was in fact cheaper for the decorator to use them. In terms of cost and use in the late 1930s, casein paints fell between distemper at the lower end of the market and oleoresinous flat paints at the higher end (Warth and Browne 1939: 325). Despite the many good characteristics of casein paint, it did not fare well in damp conditions and was not scrubbable. It was in response to these defects that oil emulsion paints were introduced in the late 1930s. For use in emulsion paints, the amount of casein decreased, and it stopped being the binding medium and instead acted as a stabilizer and thickener.

OIL EMULSION PAINTS

Oil emulsion paints appeared in the United Kingdom in the late 1930s and in the United States in the early 1940s. Oil-resin emulsions could comprise traditional drying oils such as linseed or tall oil, or alkyd resins in water; the latter were introduced just prior to World War II (Martin 1947: 148) and are discussed in detail in chapter 5. The oil or resin component was emulsified in water by a protective colloid, usually casein.

Oil emulsion paints were a logical development from the casein paints reinforced with oils or varnishes described above. However, they possessed far greater durability and represented a considerable step toward the goal of making a waterborne paint that was as durable as its solvent-borne counterpart. They were very popular during World War II, when rationing of linseed oil meant that it could be most effectively utilized in emulsified form (Scholz 1953).

To manufacture an oil emulsion paint, the varnish or oil component of the system was first emulsified with water. This was achieved by stirring it with water and an alkali, which broke the mixture into tiny oil-in-water bubbles. The dispersion was stabilized with a protective colloid. On application the friction of the brush caused the bubbles to break, and the oil component was spread over the surface, providing a more or less waterproof binder for the pigments ("Why Water Paint Flakes" 1948: 55). Drying occurred first by evaporation of the water, followed by oxidation of the oil or resin; these films therefore took a number of days to dry completely and were not scrubbable until thoroughly dry. Except for the very cheap grades, oil emulsions were pigmented with lithopone and zinc whites, with blanc fixe and whiting as extender pigments. Whiting could not constitute a major proportion of the pigmentation, as it could in distemper, due to its transparency in oil.

By 1938 oil emulsions were reportedly being widely used for both interior and exterior applications in the United Kingdom ("Distempers" 1938: 51), although they were seldom used in the United States until the early 1940s (Warth and Browne 1939: 344). They were initially supplied as a thick paste that could be thinned with water, or with a special "petrifying" liquid, which consisted of an oil-in-water emulsion. The petrifying liquid offered extra oil to assist in sealing the substrate and made the final film more water resistant. Some companies, including Sherwin-Williams in the United States and Sisson's Brothers in the United Kingdom, supplied the paints ready-to-use; this was to prove immensely popular in the United States especially and did much to boost the do-it-yourself market (see plate 13).

Sherwin-Williams's Kem-Tone, which was introduced in 1941, was an oil-in-water emulsion paint of this type. Kem-Tone comprised a linseed oil, and later tall-oil, binder that was emulsified with casein. It was supplied ready-to-use, covered a variety of surfaces in a single coat without the need for special primers, was easy to apply, had low odor, dried within an hour, and was easy to clean up. These properties made it especially popular with the home user, a market that was boosted further on the introduction of the Roller-Koater paint roller, which made the paint even easier to apply for inexperienced users.[2] Oil-in-water emulsions were the first truly washable waterborne paints and were the forerunners of the synthetic resin emulsion paints that came to dominate the flat wall coatings industry by the late 1940s and early 1950s (see chapter 6).

Interior and exterior paints based on natural oils, gums, and resins continued to be used until well into the 1950s. The introduction of phenolic resins in the late 1920s and the widespread availability of alkyds in the late 1940s did not immediately displace natural products but were manufactured alongside them for many years. However, by the 1960s it was rare to find an interior or exterior gloss paint that was not formulated with a synthetic resin, and flat wall paints based on oils were rendered all but obsolete on the widespread availability of vastly superior synthetic emulsion paints in the late 1940s and early 1950s.

Chapter 4
Nitrocellulose Lacquers

Nitrocellulose is a semisynthetic resin that was used in the manufacture of very-fast-drying lacquers, primarily for industrial applications. The first dipping lacquers based on nitrocellulose were introduced in the late nineteenth century, but it was not until spray lacquers were developed in the early twentieth century that their use in industry became widespread. Nitrocellulose lacquers are most commonly associated with automobile finishes, an application for which they were eminently suitable and phenomenally successful on their introduction in the mid-1920s. The success of industrial nitrocellulose lacquers can be attributed largely to their exceptionally fast drying time, and it was this property that encouraged manufacturers to experiment with their use for other applications, including household coatings. Although seldom mentioned in today's literature on decorative house paints, interior lacquers based on nitrocellulose were available and enjoyed a brief period of popularity in the mid- to late 1920s and continued to be manufactured until well into the 1940s. These lacquers were available for spray or brush application.

Nitrocellulose lacquers are fundamentally different from traditional oleoresinous paints and varnishes in both behavior and properties. Nitrocellulose lacquers dry primarily by evaporation of the solvents and, depending on the other resins included, remain fully or partially soluble in their carrier solvents. It is in this respect that they differ most from oleoresinous paints, which dry by oxidation and, once dry, become insoluble in their carrier solvents. The solvent sensitivity of nitrocellulose lacquers that dry by evaporation alone makes it impossible to apply a second coating by brush, as any subsequent coat simply redissolves and disrupts the underlying layer. The permanent solvent sensitivity of nitrocellulose, coupled with its extremely fast drying time, was the root cause of most of the problems encountered by both the professional decorator and the amateur who attempted to use the lacquers in their brushable form.

MANUFACTURE OF NITROCELLULOSE

Nitrocellulose and cellulose acetates are technically semisynthetic resins, as they derive from cellulose, the fundamental chemical compound in all vegetable matter. When combined with nitric and sulfuric or acetic acid, cellulose yields esters known as nitrocellulose and cellulose acetate, respectively, both of which have found use as molding materials, films, artificial silk, coatings, and lacquers. Although both forms are flammable, nitrocellulose is far more so than acetate. Indeed, the explosive guncotton is a form of highly nitrated cellulose. Both nitrates and acetates were formulated into dope, syrup used during and after World War I for tightening the fabric wings of aircraft. Both have also been formulated into coatings, although the nitrated form was used more commonly for this application.

Nitrocellulose lacquers are prepared by washing and dehydrating cellulose, reacting it with nitric acid, and dissolving the resulting esters in organic solvents. Although nitrated cellulose is capable of forming a clear coating, its solubility is poor, and it requires considerable quantities of solvents. Moreover, once the solvents have evaporated, the remaining coating is thin and brittle, and numerous coats are required to build films of sufficient thickness. Therefore, in order to formulate a serviceable coating, a gum or resin, or a mixture of the two, is included to increase solids content and gloss and promote good adhesion. Plasticizers are also added to make the coating less brittle and to confer enough flexibility to enable it to withstand expansion and contraction on weathering. The pigments, nitrocellulose, gums, resins, and plasticizers make up the film-forming part of the coating, while the solvents and diluents evaporate from the film, often with great speed.

For a decorative lacquer, alcohol-soluble nitrocellulose was chosen. This allowed less odorous solvents to be used than one might encounter for industrial applications, an important consideration in coatings for domestic use. The gums and resins also needed to be soluble in alcohol, and early coatings included natural resins such as shellac, rosin, and dewaxed dammar, and later ones ester gum and alkyd resins (Ward 1926). A mixture of solvents was necessary to ensure that all the components of the system were fully dissolved. The solvents were carefully chosen to allow the lacquer to be freely sprayed or brushed and to control the rate of evaporation to an appropriate level. The solvents did not generally affect the strength of the final film, but they did control the flow of lacquer and therefore had some influence on film thickness and properties. Finally, diluents were included; these did not necessarily have solvent action, in that they did not dissolve the component parts of the lacquer, but they reduced the viscosity and thus lowered the cost ("Why Good Lacquers" 1946: 47). Common plasticizers were phosphate, tartrate, and phthalate esters. However, the most effective plasticizers were alkyd resins, which were used for this purpose shortly after they became commercially available in 1926.

Lacquers intended for brush or spray application shared similar formulations, in that they both comprised nitrocellulose, plasticizer, gums or resins, and solvents. However, those intended for brushing contained higher solids content and slower evaporating solvents; the former ensured that sufficient film thickness could be built up with fewer coats; the latter made it possible—theoretically at least—to brush the coating out before the solvent evaporated. As with other decorative finishes, the quality of nitrocellulose lacquers varied enormously, depending on the quality of the individual ingredients, the proportions in which each was used, and the skill with which they were compounded. Prices varied considerably, and consumers generally got what they paid for.

The lower-priced lacquers tended to contain higher proportions of solvents and diluents and lower concentrations of resin and plasticizer. In some instances, the film-forming ingredients comprised as little as 10% of the lacquer, with consequential lack of build and durability. If plasticizer was not present in sufficient quantities, the film would crack and chip away more easily, but if present in too great a quantity, it imparted softness to the film. Low-cost coatings contained cheap gums and resins, and although the inclusion of large amounts of cheap resin increased the gloss and build of the film, as it was not possible to incorporate the necessary amount of plasticizer, the film could lack elasticity. The quantities of plasticizer required were so great that the hardness of the film was compromised.

By contrast, high-grade lacquers contained up to around 35% solids: the higher proportion of nitrocellulose in the film made it stronger and harder. Although less resin was included, the high-quality type chosen conferred depth and richness to the film. Greater amounts of plasticizer could be incorporated without the risk of softening the film. Finally, high-grade solvents and diluents were included. Although these evaporated from the film, they had considerable influence on the ease of application and quality of finish produced and minimized faults such as blushing, orange peel, and pin holing ("Why Good Lacquers" 1946: 47).

HISTORY OF NITROCELLULOSE COATINGS

The first coatings based on nitrocellulose were introduced by the explosives industry around 1905, as most explosives manufacturers utilized the by-products of guncotton manufacture by selling nitrocellulose dipping lacquers. Dipping lacquers were very thin and volatile, with insufficient solids content to be applied by spray or brush. They were used for coating articles such as brush handles, doorknobs, and similar small items.

The real impetus for the development of nitrocellulose coatings came in 1918, at the end of World War I. Numerous manufacturers found themselves with huge stockpiles of raw materials and plants capable of large-scale explosives manufacture but no market for their product in peacetime. This prompted companies to put their products to other uses, and coatings provided a promising alternative market. In the United Kingdom the Nobel Explosives Company devised a method of pigmenting airplane dope with the relatively new pigment, antimony oxide, and around 1919 introduced the first interior enamel based on nitrocellulose, Glossy White S.2567; it reportedly surprised DuPont's chemists as late as 1925 with its outstanding qualities ("Story of Paints Division" n.d.: 5). And in 1920 the U.K. company Frederick Crane introduced its range of Cranco nitrocellulose enamels.

It was from these early enamels that the idea for automobile finishes based on nitrocellulose originated. Cranco enamels were reportedly seen by Sir Herbert Austin, manufacturer of Austin Cars, at the 1921 British Industries Fair. He immediately recognized their potential as coatings for his cars and requested that the factory begin research into their development with this application in mind. At a similar time in the United States, the car manufacturer Henry Ford was also in search of a quick-drying coating suitable for finishing his cars. Having been assembled using the latest mass production techniques, cars coming off the production line were held stockpiled for many weeks while the traditional oil-based finishes, which took several weeks to complete, were applied. Ford made a similar request to DuPont to develop a quick-drying coating to solve this problem, and both companies began work on a nitrocellulose coating that had sufficiently low viscosity but high solids content to enable it to be applied with a spray gun.

It was DuPont who was the first to introduce a marketable nitrocellulose automobile finish. DuPont had been manufacturing a clear nitrocellulose or pyroxylin lacquer finish for furniture since 1921,[1] and throughout the summer of 1923 scientists worked with chemists and engineers from General Motors on the development of these lacquers. Duco auto paint was introduced to the United States in 1923 (Zintl 1947: 48). Duco was introduced to the British market in 1925, but the import duty made it prohibitively expensive. Therefore, DuPont set up a joint company with Nobel Industries (in 1926 merged to form ICI), called Nobel Chemical Finishes (NCF), which would manufacture and sell Duco-type finishes to Britain and her territories. Duco's formula was changed to suit the climate and available raw materials in the United Kingdom; it was renamed Belco and launched in 1926.

Following the phenomenal success of industrial coatings based on nitrocellulose lacquers, companies in both the United States and the United Kingdom were quick to recognize their potential as decorative coatings, and by the mid-1920s many firms were marketing nitrocellulose brushing lacquers for finishing metal, wooden trim, and household items such as furniture. According to *Paint Oil and Chemical Review,* by 1925 at least four large U.S. manufacturers were making brushable lacquers for household use (*The Decorator,* August 22, 1925), and numerous other companies in both the United States and the United Kingdom quickly followed suit: DuPont launched "Brush" Duco in March 1926 (see plate 14), and ICI launched "Brush" Belco in the spring of 1927 ("Ideal Home" 1928: 674). By 1930 about 250 companies in the United States were manufacturing nitrocellulose brushing lacquers intended for both industrial and domestic applications (Hounshell and Smith 1988: 145). Brushing lacquers were geared primarily to the home user, but the lacquers were also available in spray form, intended for the professional decorator.

The poor properties of brushing lacquers meant that they enjoyed only brief popularity, and by the late 1920s they had largely been superseded by alternative quick-drying products containing tung oil, ester gum, or phenolic resins. Although the market share for spray coatings was small, they were slightly more successful and were still being marketed in the late 1940s.

Plate 8

John Hall & Sons' advertisement for Brolac was one of the first in *The Decorator* to explicitly advertise tung (wood) oil in a paint's formulation. That it also refers to a "special hardening medium" suggests the presence of phenolic resin as well. *The Decorator,* October 15, 1935, 41. © V&A Images/Victoria and Albert Museum, London

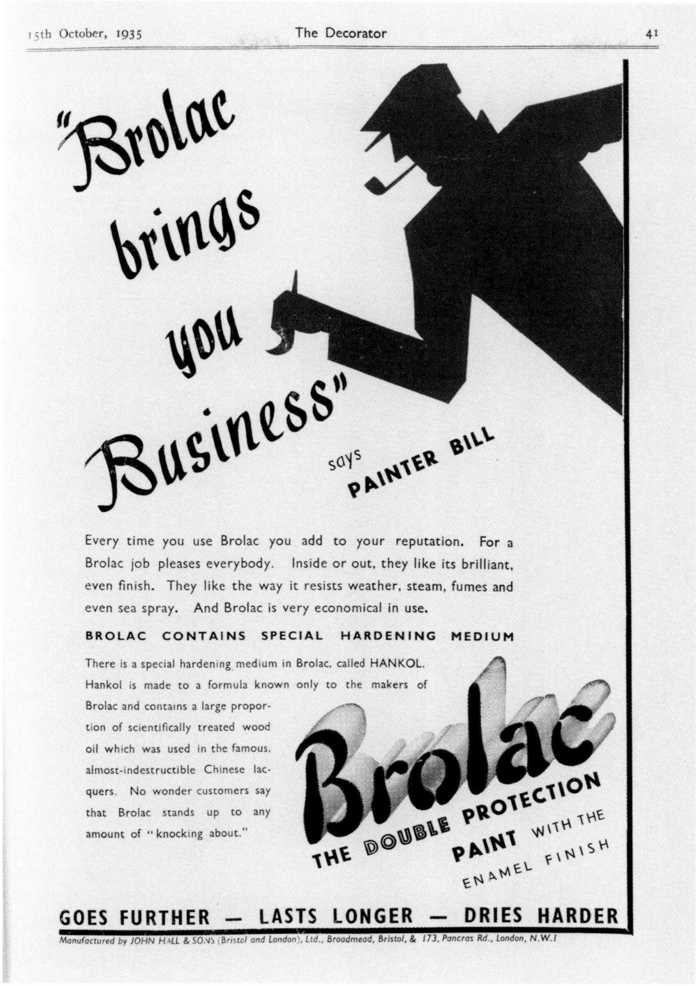

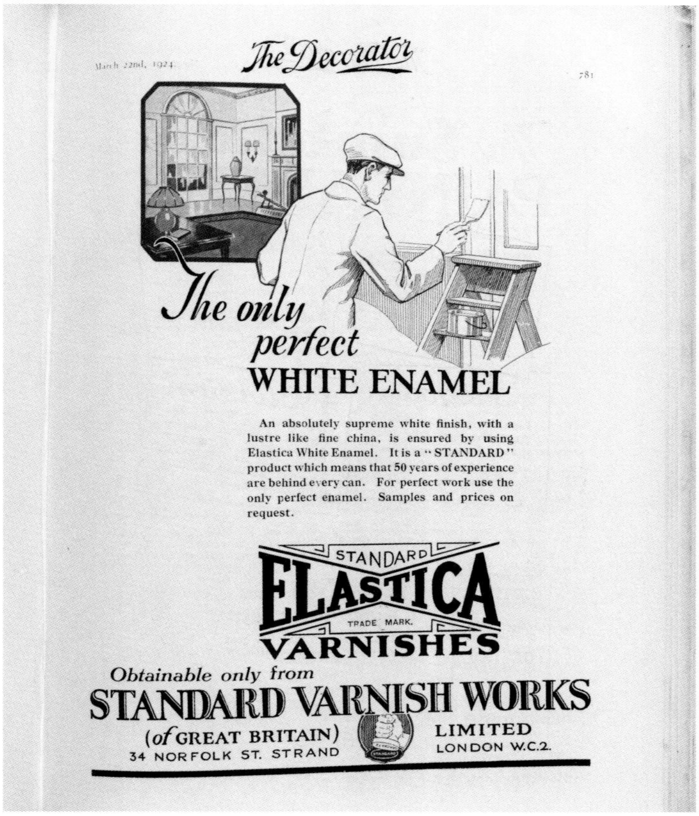

Plate 9

Standard Varnish Works Limited's Elastica White Enamel was one of many high-grade enamels that were advertised in the 1920s. Such a finish comprised zinc oxide pigment ground into a pale stand-oil vehicle. The finish was exceptionally durable and bright white in appearance. *The Decorator,* March 22, 1924, 781. © V&A Images/Victoria and Albert Museum, London

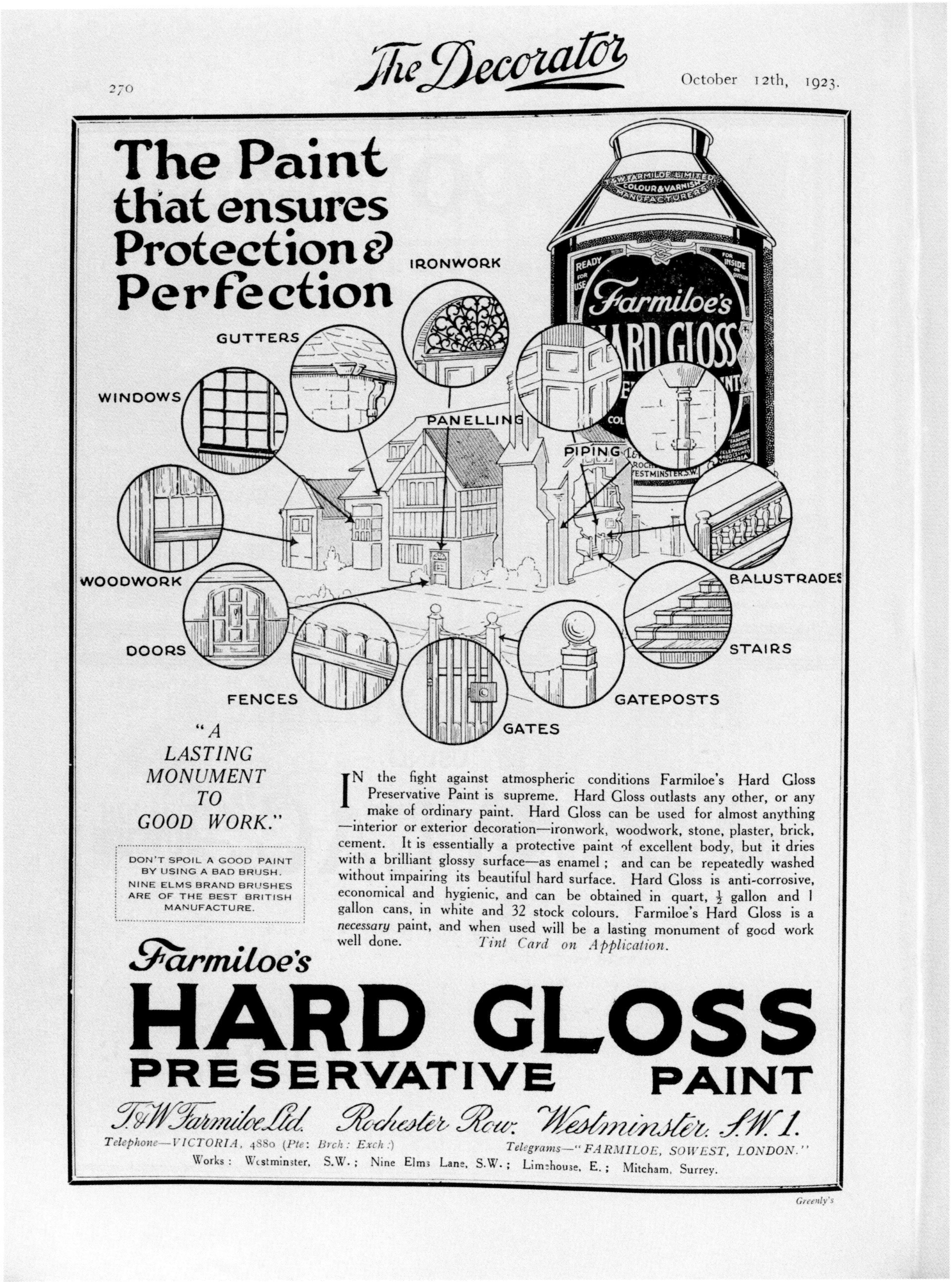

Plate 10

T. & W. Farmiloe Ltd.'s Hard Gloss Preservative Paint was marketed for exterior application, suggesting that it comprised a high-grade oleoresinous varnish and may have contained tung oil. *The Decorator,* October 12, 1923, 270. © V&A Images/Victoria and Albert Museum, London

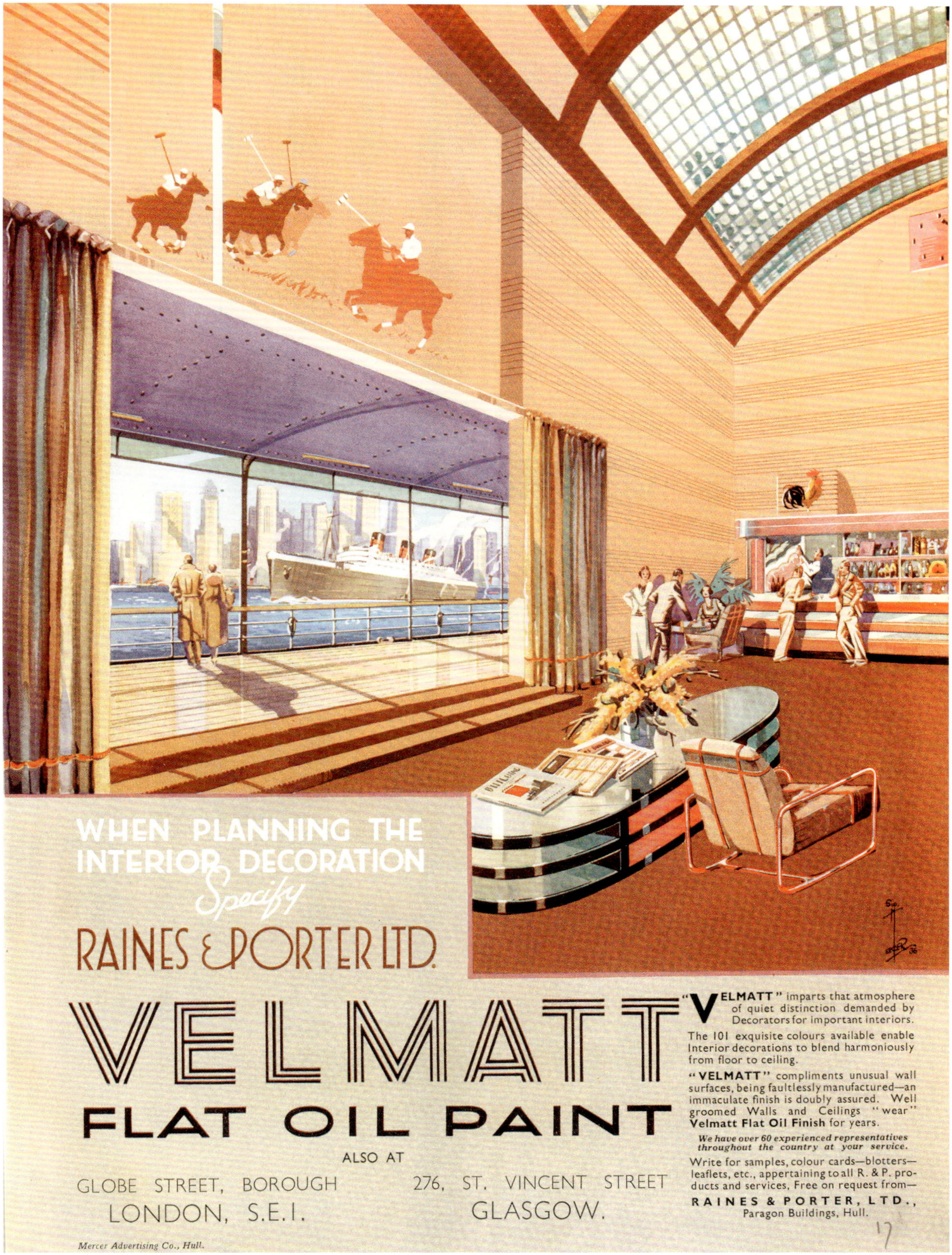

Plate 11

Raines & Porter Ltd.'s Velmatt, a flat oil paint. This was the first full-color advertisement to appear in *The Decorator*. *The Decorator,* July 5, 1936, 17.

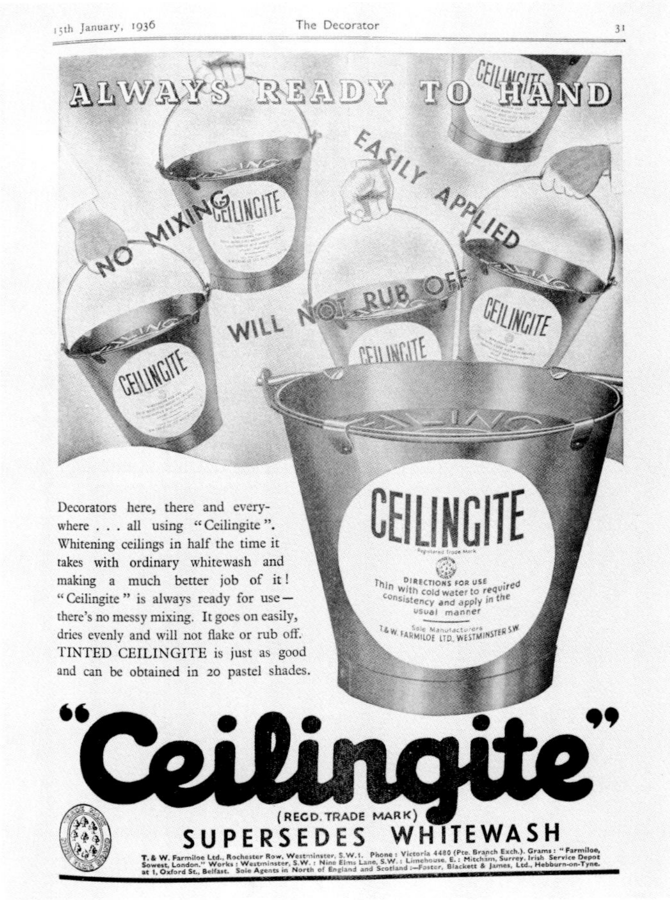

Plate 12
T. & W. Farmiloe Ltd.'s Ceilingite, a distemper formulated especially for use on ceilings. It is advertised as superior only to whitewash but is not advertised as washable, suggesting that it is an unmodified distemper. *The Decorator,* January 15, 1936, 31. © V&A Images/Victoria and Albert Museum, London

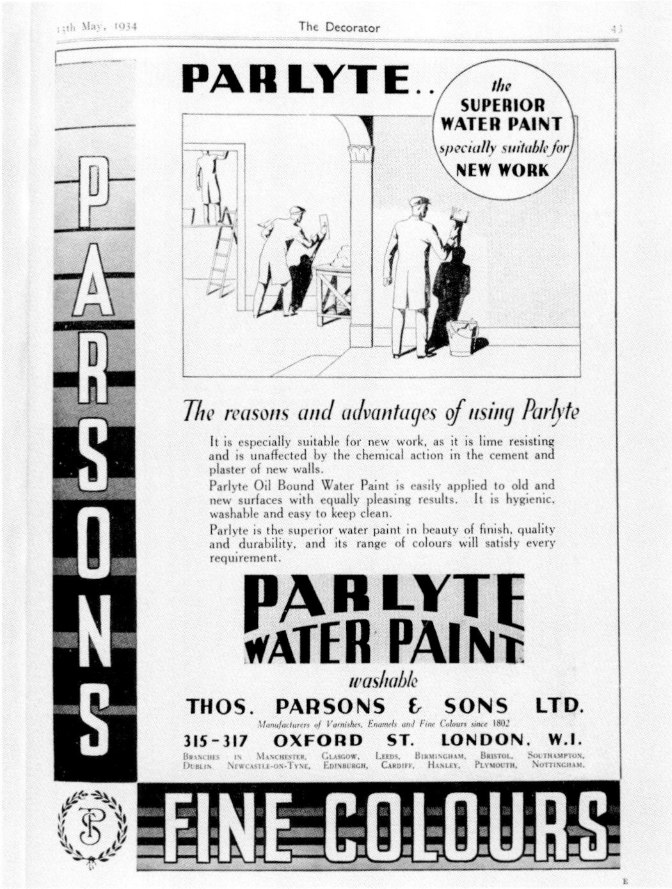

Plate 13
Advertisement for Thos. Parsons & Sons Ltd.'s Parlyte Water Paint, an oil-bound water paint for interior or exterior use. It was advertised as alkali resistant, suggesting that it contained tung oil, which had superior alkali resistance to linseed. *The Decorator,* May 15, 1934, 43. © V&A Images/Victoria and Albert Museum, London

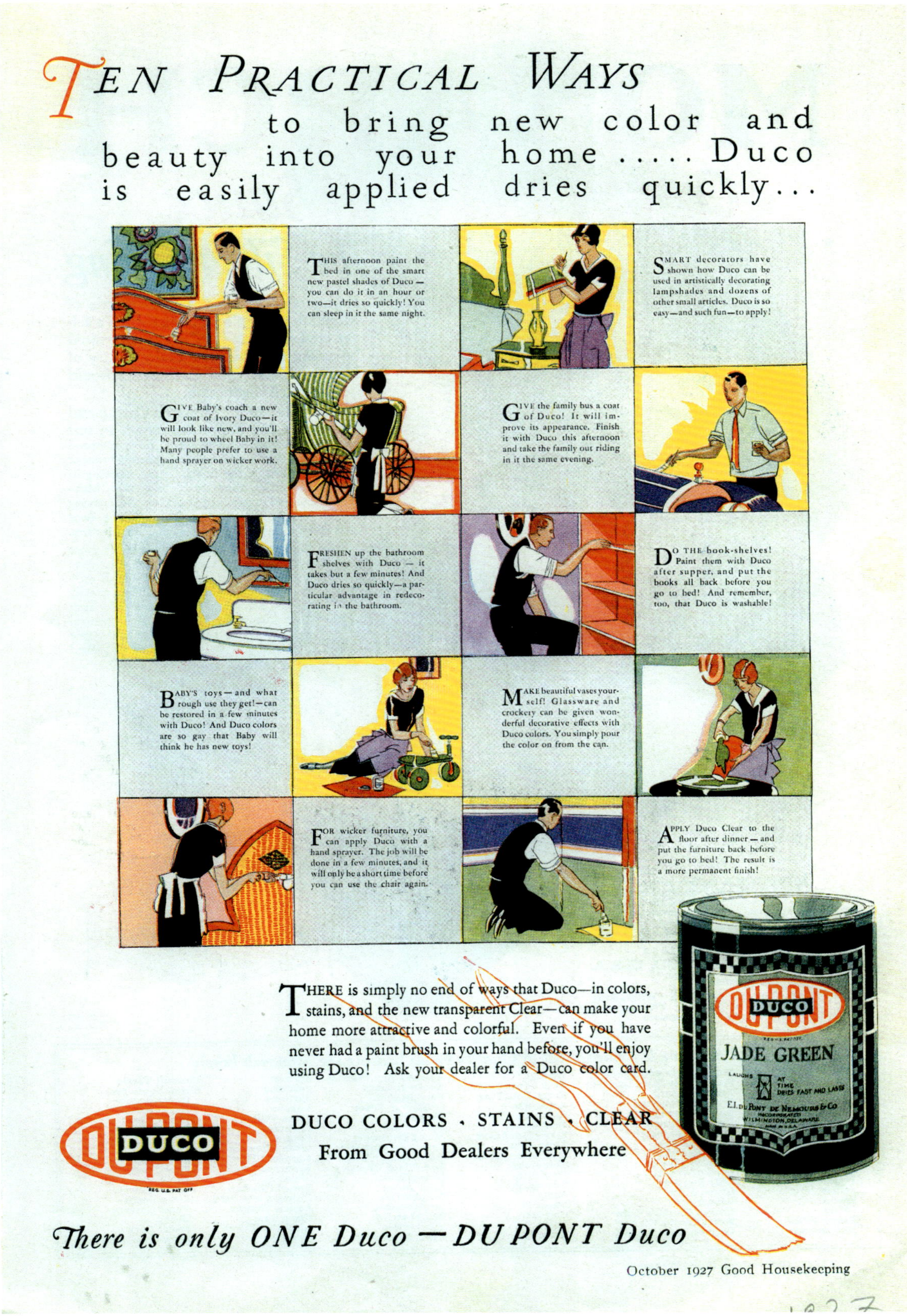

Plate 14

Advertisement for DuPont's "Brush" Duco that appeared in the October 1927 issue of *Good Housekeeping.* In the United States in particular these brushing lacquers were geared to the home user rather than the professional. Courtesy Hagley Museum and Library (P20091106_001)

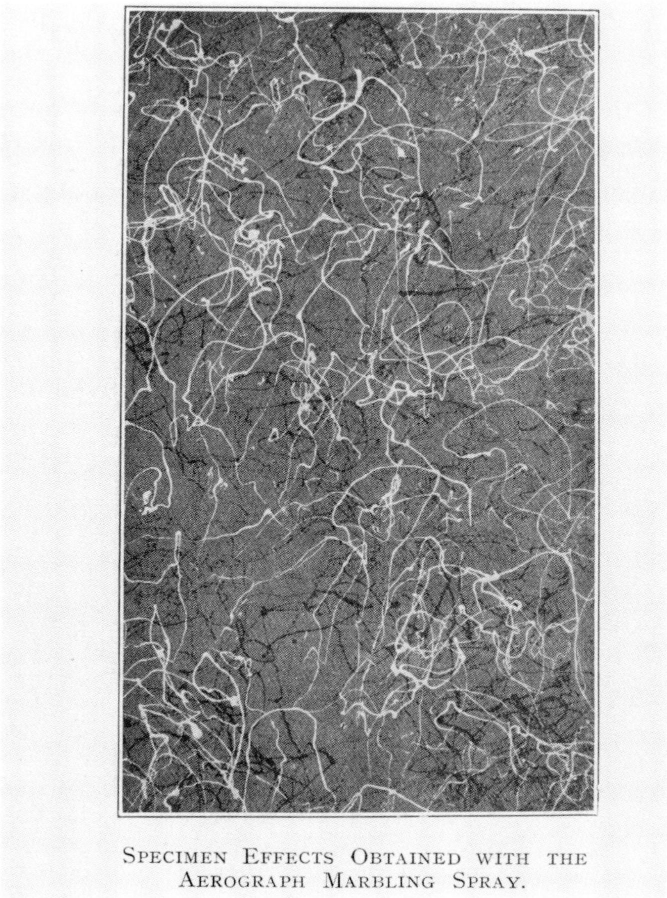

Plate 15
"Specimen Effects Obtained with the Aerograph Marbling Spray." Aerograph was one of the largest manufacturers of portable spray equipment, advertisements for which appeared with increasing regularity throughout the 1920s. The low viscosity of nitrocellulose made it especially suitable for application by spray, and it would appear that professional painters recognized the expressive potential of spray techniques with this material some time before artists did. *The Decorator,* August 23, 1926, 109. © V&A Images/Victoria and Albert Museum, London

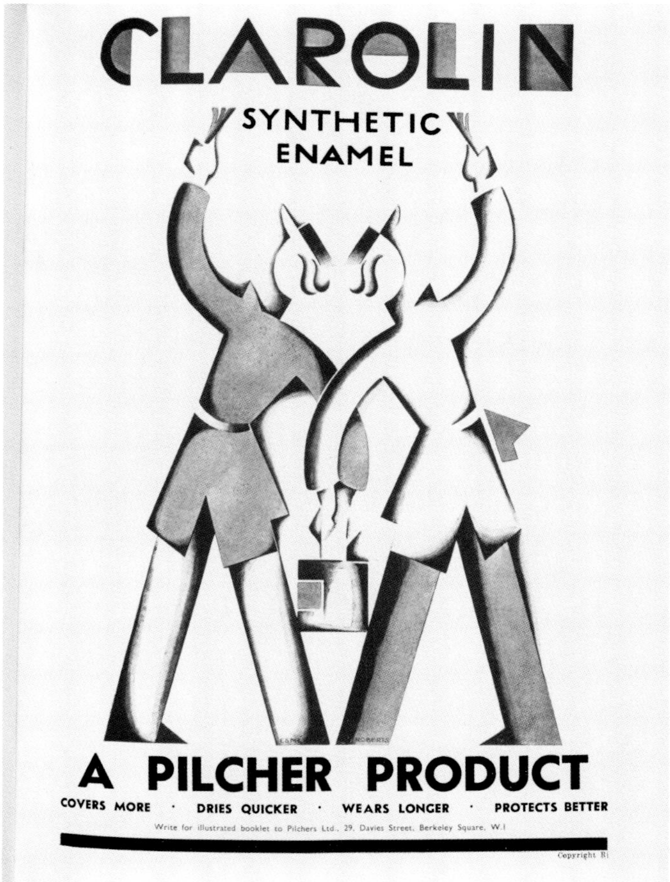

Plate 16
Pilchers Ltd.'s Clarolin Synthetic Enamel. Pilchers was one of the few companies to advertise that its products contained a synthetic resin. This paint is almost certainly based on a phenolic resin, as the use of alkyds was very limited at that time. *The Decorator,* March 15, 1935, 176. © V&A Images/Victoria and Albert Museum, London

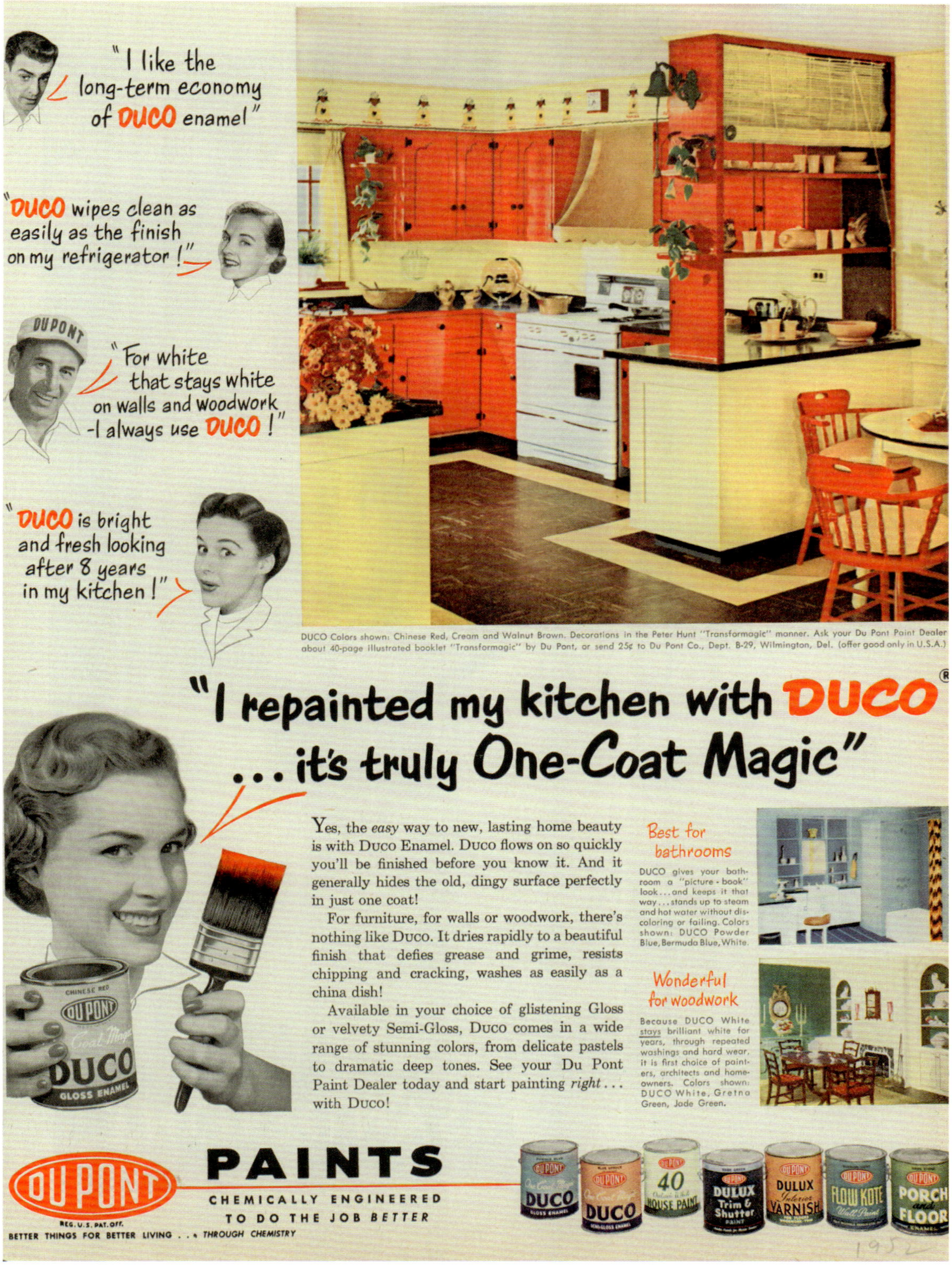

Plate 17

Print advertisement for DuPont's Duco, from *Better Homes & Gardens,* September 1952. The original brushing Duco, which was introduced in the mid-1920s, was formulated with nitrocellulose. When application problems occurred, the binder was quickly replaced with an alternative quick-drying product, and in 1934 it was reformulated once more with an alkyd resin. In 1952 Duco and Dulux were very similar products in that both were formulated with an alkyd binder, but the strength of the Duco brand is indicated by the fact that it remained a distinct product, aimed at the home user for coating interior woodwork, walls, and furniture. Acc. No. MS1083 Series I. Courtesy Hagley Museum and Library (P20100811_003)

Plate 18

Print advertisement for DuPont's Flow Kote, from the *Saturday Evening Post,* April 19, 1952. DuPont introduced its first emulsion or latex paint, Flow Kote, in April 1952. It was based on styrene-butadiene copolymer. Acc. No. 1803 Series I. Courtesy Hagley Museum and Library (P20100811_006)

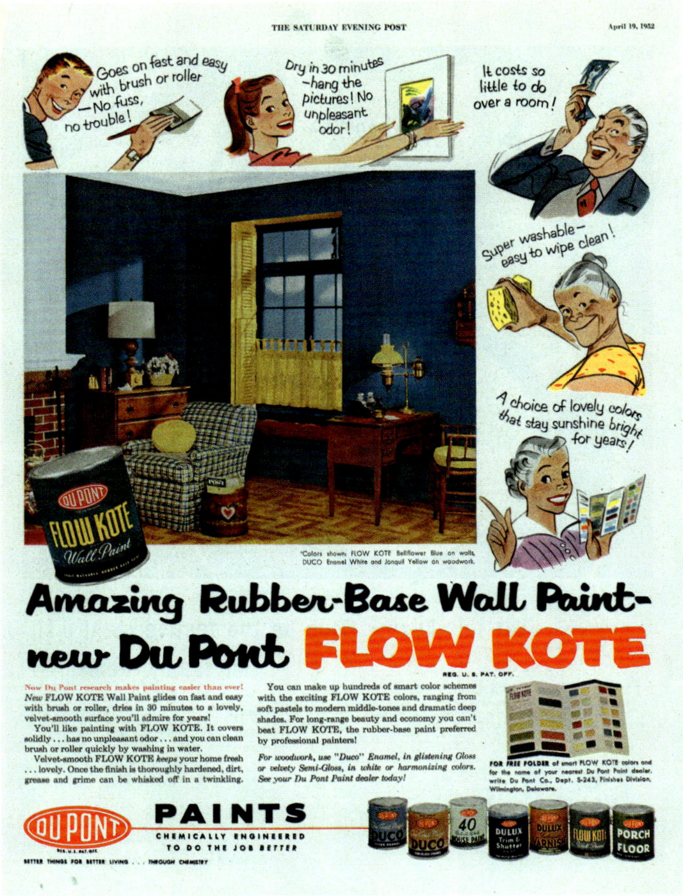

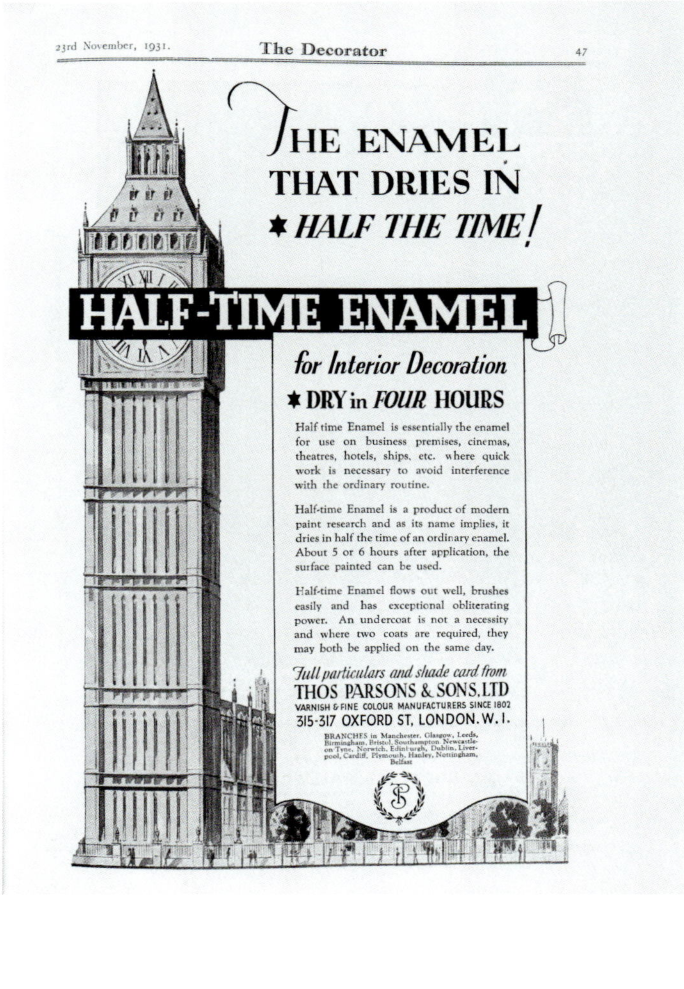

Plate 19

Thos. Parsons & Sons, Ltd.'s Half-Time Enamel, advertised as drying in four hours. This is the first advertisement in *The Decorator* to market quick-drying enamel paints. The company does not state what the vehicle is, but it almost certainly contains tung and possibly phenolic resin. The reluctance to advertise the paint vehicle was a peculiarly British trait. *The Decorator,* November 23, 1931, 47. © V&A Images/Victoria and Albert Museum, London

PROPERTIES OF NITROCELLULOSE BRUSHING LACQUERS

The main appeal of nitrocellulose lacquers lay in their extremely rapid drying time, which could be anywhere from a few minutes for industrial lacquers to somewhere between thirty minutes and two hours for brush applications. When used for industrial finishes, nitrocellulose was a phenomenally successful binder: its rapid drying time and good durability when compared to oleoresinous finishes far outweighed its disadvantages, which included the need for many coats to build up sufficient film thickness, and solvent sensitivity (Kienle 1949). However, nitrocellulose was so unsuited to brush application that the popularity of brushing lacquers is surprising; but their very fast drying time was clearly a novelty, while their excellent industrial performance helped boost their reputation—initially at least.

Many of the problems associated with nitrocellulose brushing lacquers stemmed from two factors: their high solvent content, and their fundamental difference from traditional oleoresinous finishes. The high proportion of solvents not only made lacquers expensive and unpleasant to use, but the rate at which they evaporated from the film made the coating all but impossible to brush out. The application of nitrocellulose lacquers was substantially different from both traditional water- and oil-based finishes and in fact shared more with enameling and coach painting techniques. Whereas traditional oil paints required extensive brushing out, and the slow-drying nature of linseed oil allowed plenty of time to do this, lacquers had to be applied liberally, quickly, and lightly. They could not be brushed out, laid off, or crossed—as oleoresinous paints must be—or an uneven, streaky coating resulted ("Brushing Cellulose" 1938). Indeed, the speed at which the early lacquers dried—within thirty minutes—must have been alarming for a housepainter accustomed to slow-drying oil paint.

Within minutes of application much of the solvent had evaporated from the film, and the surface became tacky. Furthermore, once the coating had dried, the nitrocellulose remained partially or wholly soluble in the carrier solvents, so attempts to patch-repair or apply a second coat ended in a sticky mess; one author complained that the application of a second coat left a surface sticky enough to pull bristles from the brush (Sproxton 1925). Indeed, special brushes also had to be employed, as the adhesive used in traditional brush manufacture was softened by the solvents in the lacquer. The problem of shedding bristles prompted one author to comment that "the time occupied in picking a bristle from a sticky lacquer surface is often sufficient to spoil the appearance of an otherwise good job" (Oliver 1928: 747–48). Moreover, it was so difficult to completely remove all traces of color from brushes that the decorator was advised to keep separate brushes for each color. There was, however, one major advantage to the solvent solubility of the lacquer: the brushes never clogged up as they did with oil paints. The application of brushing lacquers was so different from traditional products that one author felt compelled to dispense the following advice: "Resolutely banish from [your] mind all preconceived notions derived from the application of ordinary linseed oil paints or varnishes" (Sproxton 1925: 190–91).

Reluctant to take the blame for the poor performance of "Brush" Duco, DuPont blamed the women toward whom the product was geared, lamenting the company's inability to "change the habits of women who believed that they were painters[,] . . . [who had a] habit of *brushing* out lacquer like they were accustomed to doing with an oil paint, which they could brush and brush and brush" (Zintl 1947: 85–86). This was disingenuous, to say the least, as it is clear that most of the problems lay with the lacquer rather than the practitioner. DuPont did in fact admit that some of the serious faults inherent in "Brush" Duco were due to the fact that, in response to competitors' products, the paint had been rushed onto the market without adequate testing. Aware that the poor quality of "Brush" Duco might have an impact on the excellent reputation of the industrial version, the product was changed in the can to a conventional quick-drying product.

Although the faults described above were clearly inherent in the lacquer, the professional painter had to take some of the blame for the poor performance of brushing lacquers. Lacquers were so different from traditional oleoresinous paints and varnishes that it is perhaps not surprising that users were baffled by their properties and correct use. Those who were accustomed to mixing their own paints and varnishes—adding extra oil, driers, turpentine, and pigments as the finish warranted—were suddenly confronted with a completely different type of finish that was incompatible with the solvents and oils traditionally used by them and could not be adulterated without disastrous results.

The painter could not add dry pigment if he wished to alter the color, as was possible with traditional paints, but needed to mix two lacquers together. Furthermore, it was advisable that the mixed lacquers be from the same line of paints: the latitude of materials used in manufacture was so great that many brands were incompatible. If the painter wished to thin the paint, it had to be done with special thinners, again from the same manufacturer as the paint: unfortunately, the painter sometimes interpreted manufacturers' insistence on buying its branded thinners as a barely concealed attempt at conning him into buying unnecessary and expensive solvents, so the advice was often ignored. That the professional painter had a tendency to adulterate these finishes inappropriately, to the extent that they could be rendered useless, is evidenced by the British company Naylor Brothers' recognition of the need to educate the painter in the proper use of these materials in a 1928 conference ("Cellulose Paints" 1928).

Despite these considerable shortcomings, brushing lacquers were initially popular, especially in the United States, which can largely be attributed to the fact that they were marketed to the female home user for the decoration of small household items such as furniture. These were one-time users who were attracted by the rapid drying time and were more likely to be swayed by the massive advertising campaigns launched by the large companies that manufactured the paints. It seems that few professional painters persevered with nitrocellulose brushing lacquers, as they were so unsuited to brush application. Indeed, as early as 1926 the trade press noted that other than when extremely rapid drying was required, there was no benefit of nitrocellulose lacquers over traditional paints and varnishes (Ward 1926).

However, for the professional painter who was willing to invest time and money in learning a new technique, spray nitrocellulose wall paints were a viable option, and the popularity of these coatings for some applications is indicated by the increasing number of manufacturers advertising portable spray equipment throughout the late 1920s, 1930s, and 1940s. By applying with a spray gun, it was possible to take advantage of the good properties of nitrocellulose while minimizing the poor ones.

PROPERTIES OF NITROCELLULOSE SPRAYING LACQUERS

The advantages of spray applications were clear: when applied by spray, a typical room could be coated in just twenty minutes and recoated ten minutes later ("A New Method" 1927: 740). Moreover, the fact it was applied by spray meant that numerous coats could be applied without disrupting the previous ones. The coating dried so quickly that it was especially suited for painting stairs, schools, hospitals, and hotels, as rooms could be occupied as soon as the unpleasant odor from the solvents had dispersed. There were other advantages too: the binder contained no oil, so it could not be saponified by the alkaline nature of new plaster. This offered an immediate advantage over both solvent- and water-based paints containing oil, which would be destroyed in such conditions. It did, however, mean that all the pigments included needed to be alkali resistant ("Cellulose" 1946: 60). Nitrocellulose could be applied to most surfaces, other than mahogany-stained wood and fresh paint, although an undercoat was recommended if the paint was applied to bare metal or if the existing coating was badly chipped ("Cellulose Paints" 1928).

The binder itself was colorless, so it could be tinted with delicate colors ("Cellulose" 1934: 32), another advantage over oil paints, which had a tendency to yellow. Nitrocellulose was also impervious to moisture, which could be both a benefit and a problem: the finished coating was highly washable, but if the plaster was not perfectly dry before application, the moisture trapped beneath caused severe blistering and flaking of the coating. Although the initial cost of nitrocellulose lacquers was appreciably higher than oil or water paints, their incredibly fast drying time offset both the labor costs and the costs associated with closure of buildings while coatings dried.

Although their advantages were clear, there were numerous drawbacks to using spray nitrocellulose decorative finishes, including the need for specialist equipment, the investment in learning a completely new application technique, the strong smell, and the health consequences of using some fairly noxious solvents. The last point appears to have been of little concern in the 1920s, when health and safety advice was largely limited to warning against lit cigarettes near the lacquers ("Brushing Belco" 1928: 456). Indeed, in 1928 a Mr. Wilson, who manufactured Belco finishes, made the somewhat dubious claim that his workers' health had actually improved since being around cellulose lacquers, and aside from "a tickle in the throat on the first day," he himself was in better health than he had ever been ("Cellulose Paints" 1928: 376–82). However, Wilson clearly had his doubts: although he was willing to issue a statement to the effect that the lacquers were harmless, the possibility of legal expenses prevented him from providing a guarantee.

The smell of the early nitrocellulose finishes derived from the use of amyl acetate and *n*-butyl acetate, solvents with a strong odor of pear drops or banana. When the paint was applied, the smell tended to permeate everything in the home, and although the "ladies" at first reportedly rather liked it, it quickly became "nauseating" and was certainly a factor that counted against nitrocellulose finishes (*The Decorator,* April 22, 1930, 10). By 1932 the danger of exposure to cellulose solvents was at last recognized: one manufacturer conducted an examination, including blood tests, of 133 operatives who had complained of irritation, headache, depression, and fatigue after long exposure (*The Decorator,* July 22, 1932, 30).

Another major disadvantage of nitrocellulose was that it was, of course, highly flammable. Early versions of lacquers had a flashpoint of over 73°F, rendering them subject to petroleum restrictions for storage. This was a significant handicap for people wishing to store paint and make it available for immediate use, and it could also be a problem for manufacturers and stockists: not least because of the attendant fire hazard. However, by 1928 the flashpoint of most lacquers had been lowered, and petroleum restrictions no longer applied ("Cellulose Paints" 1928).

AVAILABILITY AND USE OF DECORATIVE NITROCELLULOSE LACQUERS

Nitrocellulose was certainly phenomenally successful as an industrial finish: by 1926 nearly three million automobiles, millions of dollars' worth of furniture, railway cars, day coaches, pianos, and tank cars in the United States alone reportedly used it ("Nitro-Cellulose" 1926). Therefore, when manufacturers first introduced a range of brushing lacquers based on nitrocellulose around 1925, the same excellent characteristics were expected of them. They were thus met with great acclaim, "hailed as an epoch-making discovery which was going to revolutionise the painting industry" (Jensen & Nicholson Ltd. 1930).

Brushing lacquers were initially hugely popular, as evidenced by the number of U.S. companies manufacturing them, which increased from about 4 in 1925 to more than 250 by 1929. They were more widely manufactured and used in the United States than the United Kingdom, no doubt due to the far larger retail market in the former country, whose companies had long recognized the potential of marketing their products to the amateur painter. It very quickly became apparent, however, that early lacquers were replete with faults, including very low solids content, extreme solvent sensitivity, and a drying time that was far too rapid for brush application. Further damaging their reputation were the enthusiastic claims that they could be applied by anyone—which indeed they could, although not necessarily with good results (Cruickshank Smith 1928). Despite the early success of lacquers, their faults made them unpopular with both dealers and users. A survey of their use carried out in 1929 by the *Saturday Evening Post* in the United States revealed that 68% of dealers who carried a lacquer line were not satisfied with it, with most complaints concerning the odor (Richter 1929).

Brushing lacquers thus fell quickly out of favor with professional decorators and never really recovered from the damaging impact of the early faults. Although several attempts were made to place improved brushing lacquers on the market in the late 1920s, none reportedly satisfied the house decorator, and paint still held the dominant position (*The Decorator,* April 22, 1930, 10). In an effort to overcome the shortcomings of brushing lacquers, manufacturers in the United States in particular worked hard to improve their properties and attempted to market them to the professional rather than amateur user. Much of the research involved increasing the solids content and reducing the solvent sensitivity of the finish, which they achieved by introducing oxidizing resins. This not only improved the brushing properties, gloss, and build but also enabled the finish to be recoated, as drying no longer occurred by evaporation alone. But despite these improvements, their properties in this respect were still inferior to oleoresinous coatings based on synthetic resins ("Varnish or Lacquer" 1950: 111), and they were eventually superseded by them.

However, when applied by spray, decorative nitrocellulose paints had many good qualities, and they certainly enjoyed more success in this field. In the United Kingdom the first residence claimed to be entirely decorated using this technique was in the spring of 1927, on a new house situated at Grange-over-Sands, overlooking Morecombe Bay ("A New Method" 1927: 740). The walls first received a coat of water paint, then a spray application of nitrocellulose. Each room took no more than twenty minutes to paint and could be recoated in just ten: the speed of application convinced the decorators, Messrs J. M. Smith and Son, of Liverpool, that all home owners would soon be demanding these types of finishes. Although their use in the home never really took off, they were still recognized as useful finishes for large buildings and were still enjoying a comparatively small market share in the late 1940s ("Cellulose" 1946: 60; "Applying" 1949: 85).

Chapter 5
Oleoresinous Paints Based on Synthetic Resins

Many of the synthetic resins developed by the plastics industry in the late nineteenth and early twentieth century could be formulated into coatings, and in the early decades of the twentieth century synthetic resins began to replace natural products for this application. Some synthetic resins were used as replacements for natural resins in oleoresinous paints and varnishes, whereas others were emulsified in water (see chapter 6 for waterborne coatings based on synthetic resins). The two primary resins used in the manufacture of oleoresinous house paints based on synthetic resins were phenol-formaldehyde and alkyds, although of the two, alkyds were and still are by far the most important binder.

Synthetic resins have a number of advantages over natural types: they not only have batch-to-batch reproducibility but also usually have better color, improved durability, and confer quick-drying properties to the paint without detriment to the longevity of the film. The drying time of traditional oleoresinous finishes had been achieved by the addition of extra driers and/or natural resins, both of which shortened the life of the film considerably.

PHENOL-FORMALDEHYDE

Phenol-formaldehyde, or phenolic, resins are most commonly known in the form of Bakelite plastic. They are manufactured by a condensation reaction between phenols and aldehydes, and the resulting resin is reacted with oil to make paints and varnishes. Phenolics were used as a direct replacement for fossil resins in oleoresinous coatings, and finishes containing them were quick drying with excellent water, alkali, and chemical resistance. Since their widespread availability in the late 1920s they have been used for industrial and marine finishes, and throughout the 1930s they formed the basis of many decorative four-hour enamels: quick-drying paints that were used for finishing wood and metal and were also available in gloss and flat finishes for interior trim and walls. Phenolics were superseded by alkyd resins in oleoresinous household paints when synthetic resins became available again after World War II. Phenolics were also a useful addition to alkyd resins, where they increased drying time and water and chemical resistance. They are still widely manufactured today for numerous industrial applications but are no longer used in decorative coatings.

Manufacture of Phenol-formaldehyde Resins

Phenolic resins are made by a condensation reaction between phenols and aldehydes with an acid or alkaline catalyst. The parent compounds, phenol and formaldehyde, are undoubtedly the most important intermediates, although for use in paints, cresols and xylenols reacted with formaldehyde were also common. Although a great many different types of phenol-formaldehyde resins were available at the height of their use in decorative paints in the mid-1930s, the paint and varnish makers' requirements were satisfied by three or four grades, each differing in color, hardness, solubility, and acidity (Sanderson 1934: 714). Phenolic resins are soluble in alcohols, ketones, and similar solvents but are insoluble in the hydrocarbon solvents traditionally used in paint manufacture. Phenolics are also insoluble in drying oils, but oil solubility can be achieved by modification with rosin or ester gum or by the use of 100%, or "straight," phenols. The latter are manufactured by combining synthetic substituted phenols such as para-phenylphenol and para-tertiary butyl phenol, with formaldehyde. The resulting resin is directly soluble in all true drying oils, making the inclusion of rosin or ester gum unnecessary. Varnishes made from 100% phenols have excellent corrosion resistance and durability but poor lightfastness. Due to both the poor color and the expense of 100% phenolics, only modified types were used in the manufacture of household paints.

Phenolic resins are divided into two classes: heat hardenable or permanently fusible. Heat-hardenable resins are alkaline catalyzed and reacted with an excess of formaldehyde. They are used primarily for stoving and heat-hardening materials and were originally known as resoles. Permanently fusible resins are acid catalyzed with a lower formaldehyde ratio and are used to produce air-drying films that were originally known as Novolacs (Lilley 1943: 261–68).

Phenolics can be further subdivided into two categories: spirit soluble and oil soluble. Spirit-soluble phenolics are the oldest form of coatings and were used extensively in baking finishes. Although they offer very hard, resistant coatings, they are also extremely brittle, impossible to apply by brush, and very difficult to apply by spray. The resins used in decorative paints thus comprised acid-catalyzed, permanently fusible, oil-soluble resins, which were available in modified or 100% form (Simmonds, Ellis, and Bigelow 1943: 321–22). Modified phenolics were commonly known as Albertols, to differentiate them from ester gum.

Phenolic resins are compatible only with true drying oils, of which tung was undoubtedly the most important. The earliest modified and straight phenolics were dissolved in tung, or a combination of linseed and tung, and also perilla, oiticica, and dehydrated castor oils as they became available (for a discussion of these oils, see chapter 2). The choice of oil was important, as phenol-formaldehyde resins performed very differently with respect to linseed and tung oil. Linseed oil was not as successful as tung, and in early varnishes certainly, tung had to be present as at least part of the oil. Because phenol-formaldehyde resins were prepared at such high temperatures, the drying oil incorporated underwent chemical changes, which caused the oil to body during the process. The tendency of tung to gel at very high temperatures made it incompatible with very hard phenolic resins that needed cooking at high temperatures.

The manufacture of phenolics reportedly developed along two distinct lines in Germany and the United States, with each country employing different cooking techniques. Whereas Americans preferred to use high temperatures and a short cooking time, Germans favored low temperatures, a longer cooking time, and the use of bodied oils. However, the final products were reportedly very similar (Lilley 1943: 264–66).

A third type of resin, maleic rosin, was marketed alongside the phenolics. Although in reality maleic rosin types are closely related to alkyds, most companies marketed them under the same brand name as their phenolic types: one such example was Rohm & Haas's Amberol family of resins, which included modified phenolics, 100% phenolics, and maleic rosin types. Maleic rosin types were manufactured by heating rosin and maleic anhydride together to form a hard, brittle, resinlike material, which had a considerably higher melting point than the constituent compounds. The product was then esterified with glycerol to form a maleic-modified rosin ester. Maleic types had toughness and hardness equivalent to oil-modified phenolic varnishes ("Varnish Gums" 1950: 43) but had considerably better color retention and gloss. They were diverse resins that were incorporated into oleoresinous baking finishes and air-drying finishes and used as modifying resins for alkyds and nitrocellulose finishes. Rohm & Haas's Amberol was marketed for use in binders consisting of large quantities of soft oils and in the formulation of architectural white and four-hour enamels (Rohm & Haas 1953: 19).

A typical recipe dating from the mid-1930s describes a four-hour varnish comprising twenty to forty gallons of wood oil[1] and four to six gallons of heat-bodied linseed oil per one hundred pounds of hard or extra-hard resin. A typical manufacturing procedure was to cook the resin with the wood oil to a top heat of about 550°F, then check the mixture by the addition of linseed and sometimes a few pounds of fused lead resinate (Sanderson 1934: 714).

A varnish recipe dating from the mid-1950s describes a similar process to manufacture a general-purpose varnish for interior or exterior use: it comprised modified phenolic resin, kettle-bodied linseed oil, tung oil, mineral spirits, cobalt napthenate, and lead napthenate. The properties of the varnish could be altered by varying the components and the ratios: for example, the durability could be increased by using more phenolic, or 100% phenolic, and the cost could be lowered by substituting linseed for other oil combinations, such as oiticica, fish, or tung. The very best exterior varnishes were known as marine spar varnishes. These were extremely tough, weather resistant, and durable. A typical binder comprised 100% phenolic, modified phenolic, alkali-refined linseed oil, and tung oil. However, it was not always necessary to have such a high-performance coating, and a less expensive but adequate coating could be manufactured by using modified phenolic, varnish makers' linseed oil, and tung oil (Singer 1954a: 76–80).

History

Phenolic resins were developed by the plastics industry in the late eighteenth and early nineteenth century and were originally designed to replace the electrical insulators based on natural products. In 1872 Adolph von Baeyer conducted the first rigorous experiments that produced phenol-formaldehyde condensates, and in 1902 C. H. Mayer, working for the Louis Blumer Company in Germany, invented Laccain, the first synthetic substitute for shellac (Gardziella, Pilato, and Knop 1999: 110). The bulk of the research into heat-hardenable phenol-formaldehyde was conducted by two chemists: Sir James Swinburne in Britain and Belgian-born Leo Hendrick Baekeland in the United States (Reboul 1994: 26). Although both scientists conducted a considerable amount of research, it was Baekeland who was the first to patent a process for the manufacture of heat- and pressure-cured phenol-formaldehyde in 1907, and his Bakelite resin was introduced in 1909.

Bakelite was a hard, infusible, nonflammable resin made from spirit-soluble phenol-formaldehyde. Although developed primarily as an electrical insulator, it later found extensive use in the manufacture of cheap, mass-produced plastic consumer goods. Spirit-soluble resins were exceptionally hard and resistant to water and chemicals. Coatings could be made from them, but they had poor color and were extremely brittle and were therefore used only where extreme durability was required (Simmonds, Ellis, and Bigelow 1943: 321). But in 1910 L. Berend, working for the Dr. Kurt Albert Company in Germany, developed the first oil-soluble phenol-formaldehyde resin: oil solubility was conferred by heating the components with an excess of rosin (Gardziella, Pilato, and Knop 1999: 10). The first modified resins were introduced in Europe in about 1912.

These early resins could be manufactured by two methods: heating the phenol, formaldehyde, and rosin together or heating a previously prepared acid-catalyzed (Novolac) resin with rosin or coumarone resin; the acidity of the rosin could be reduced by esterifying the final product with glycerine (Martin 1947: 34). The components were heated at high temperatures for long periods, and the resulting resins were dark colored, with poor durability (Stoye, Freitag, and Beuschel 1996: 148). Because the resins were simply dispersions of phenol-formaldehyde in rosin, they had many of the defects associated with rosin, including softness, bad color, and poor alkali resistance. Moreover, these resins were expensive: rosin was the cheapest ingredient, but it could be present only in small quantities, as it imparted such softness to the film. Therefore, to compete with hard copal varnishes, considerable amounts of the more expensive phenol-formaldehyde had to be present.

Throughout the 1910s improvements were made to modified phenolics; much of the work was conducted by A. Amann and L. Fonrobert in Germany. They found that by using the newly available ester gum in the place of rosin, a vastly superior resin could be produced. Phenolics modified with ester gum had a very high melting point but good oil solubility. They therefore produced considerably harder films than resins formulated with rosin, and they also had greater durability and better color.

Significantly, the resin was also cheaper. Whereas rosin-modified types required large amounts of phenol-formaldehyde to produce a film with sufficient hardness, no more than 25% phenolic was required if ester gum was used; indeed, so little phenol was present that these types were sometimes referred to as "phenol-hardened ester gums" (Martin 1947: 35–38). In order to differentiate them from ester gum, ester gum–modified phenolics were commonly known as Albertols. These resins were introduced in Europe in 1916 and in the United States in the early to mid-1920s; some authors claim 1922 (Damitz, Murphy, and Mattiello 1943: 201), others 1926 (e.g., Powers 1951: 1770).

The development of phenolic resins at that time was boosted by the availability of tung oil, which increased dramatically during World War I as linseed became scarce. Although tung performed poorly with fossil resins, when combined with modified phenolics it produced excellent coatings: indeed, early phenolics could be made successfully only if the oil component was all or at least part tung, and one author contended that phenolic resins would never have been successfully made if it were not for the abundant supply of tung (Lilley 1943: 265). Further encouraging the development of phenolics was the competition from fast-drying nitrocellulose lacquers, which had stimulated demand for quick-drying products in general (Rohm & Haas 1953: 1), and in response to this competition many companies began to manufacture four-hour enamels and varnishes, many of which were based on phenolic resins.

Most of the research that enabled Albertol resins to be made into four-hour varnishes was carried out in the United States (Fonrobert et al. 1930: 88). Rohm & Haas was the first U.S. company to manufacture Albertol resins, which were introduced in 1926. In 1927 Rohm & Haas established the Resinous Products and Chemical Company, an affiliate set up specifically to exploit Albertols; one of the first things the company did was change the name of the resin to Amberol in the hope that the evocation of expensive amber would encourage manufacturers to use it. U.S. sales of modified phenolics were initially modest. The first manufacturer to make use of them in floor and interior varnishes was reportedly the George Wetherill Company, which, in 1926, introduced floor and interior varnishes based on Albertol 111L or Amberol B/S1 in combination with tung oil. Claims that the varnish dried and could be recoated in four hours proved true, and by late 1927 numerous other manufacturers began to produce their own varnishes using Amberol resins.

The Resinous Products and Chemical Company appears to have been the only company in the United States manufacturing modified phenol-formaldehyde resins for coatings during this period, and sales of them increased dramatically: from $173,000 in 1927 to $962,000 in 1928. The company claimed that by 1928 nearly every U.S. varnish manufacturer had reformulated its interior and floor varnishes with Amberol resins and that soon after, their other clear varnishes and colored enamels were similarly reformulated. By 1928 Amberol had reportedly displaced at least 90% of all other resins used in the U.S. varnish and enamel trade (Hochheiser 1986: 42–44). However, despite their huge popularity, the poor reputation of rosin and the lack of durability its derivatives conferred to Albertol coatings prompted manufacturers to develop 100%, or straight, phenols; these were directly soluble in drying oils and eliminated the need for the rosin component completely.

Straight, or 100%, phenols were developed in 1928. The work is associated with H. Hönel, the firm of Beck-Koller, and workers at the Bakelite Corporation. Direct solubility in drying oils was achieved by using highly selected or substituted phenols. The resulting resins had outstanding durability and alkali resistance, although their lightfastness was reportedly poor. These resins were initially manufactured by Beck-Koller in Britain and by the Bakelite Corporation in the United States and found use for both domestic and industrial purposes (Lilley 1943); they were especially useful in the formulation of exterior and marine varnishes. One hundred percent phenolics were intended for use in tung oil varnishes and had good solubility in oils and mineral thinners. Varnishes made from them could be diluted almost infinitely, and they also blended with ester gum; they were often used with rosin-modified resins as a fortifying agent (Rohm & Haas 1953: 13). One hundred percent phenols were more expensive than modified types, and although their properties were superior, they did not displace the latter, primarily because of their high cost.

Phenolic resins are used extensively in industrial coatings to this day, although they are no longer used for decorative paints. As with all synthetic resins, their use for trade paints ceased during World War II, and when the raw materials became available once again in the late 1940s, they were superseded by alkyd resins for this purpose.

Properties of Paints Based on Phenol-formaldehyde

Phenolic resins are very fast drying, tough, and durable. They are resistant to moisture, alkalis, acids, corrosion, and certain solvents. As discussed above, the earliest spirit-soluble types were not suited to paint and varnish manufacture, because, though very hard and with excellent corrosion, chemical, and solvent resistance, they were brittle, impossible to apply by brush, had poor adhesion to metals, lacked durability, and yellowed severely on weathering (Lilley 1943). It was not until oil-soluble types were introduced in the first decades of the twentieth century that phenolics became a useful ingredient for the paint and varnish manufacturer for the production of air-dry and baking finishes.

Phenolic resins were and are used primarily in the manufacture of industrial coatings. Their good moisture and alkali resistance made them useful for marine coatings, and they have also been used for metals, printing inks, primers, and baking finishes. The main application of these quick-drying coatings or four-hour enamels was for finishing autos, buses, and metalware (Fonrobert et al. 1930: 90), but they also formed the basis of many of the quick-drying gloss and flat decorative paints that were widely advertised throughout the 1930s.

Phenolic resins were in direct competition with very-fast-drying nitrocellulose lacquers (see chapter 4), and it was reportedly the threat from these finishes that drove much of the research into the development of phenolics (Sanderson 1934: 714). However, in terms of brush-applied paints, phenolics offered a drying speed similar to nitrocellulose lacquer but with none of its application and solvent-sensitivity problems, and they swiftly replaced nitrocellulose brushing lacquers.

From the user's perspective phenolics had many advantages over alternative finishes: their rapid drying time enabled a coating to be applied, flatted down, and recoated within six hours; they had considerably better flow and gloss than paints based on ester gum and natural copal; and they were nonsaponifiable and thus could be used in undercoats and applied to new plaster (Munns 1932: 46). Unlike nitrocellulose lacquers, phenolics incorporated the same oils and thinners as traditional paints, which meant that they had application properties similar to traditional paints and varnishes; finally, from the manufacturer's perspective, the batch-to-batch reproducibility made it easier to formulate a product with consistent properties.

However, the best property of phenolics—their rapid drying time—was also the feature that made painters in the United Kingdom in particular mistrustful of them. This was perhaps understandable, as previous attempts to accelerate the drying time of coatings had been achieved by means of the addition of either lots of resins or lots of drier, both of which were detrimental to the film's durability. It therefore took some time to convince the decorator that quick drying was not at the expense of durability.

There was, however, a significant flaw in phenolic resins, and one that was especially pertinent for decorative paints: their tendency to discolor. Phenolics generally had poor color retention and, depending on the intermediates used in their manufacture, the initial color could also be very poor. Although all phenolics had a tendency to yellow, this was especially pronounced in the para-phenylphenol resins (Molloy and Reilly 1948: 98). Using cresol in the place of phenol also yielded resins with marked yellowing tendencies. However, cresylic resins were cheaper and produced varnishes with better water resistance than phenolic types, so they might be used for applications such as undercoats and primers, where good color was not of primary importance. The best-colored products were made from rosin-modified maleic types, which had good color retention in paints and varnishes. However, Rohm & Haas, one of the largest manufacturers of phenolic resins, was unable to describe even its palest grades as anything more than "satisfactory" for producing pale varnishes (Rohm & Haas 1953: 22), and many companies did not manufacture decorative paints based on phenolics at all due to their tendency to discolor.

Availability and Use of Phenol-formaldehyde Paints

The resins used for decorative paint formulation comprised modified phenolics or rosin-modified maleic types. The use of phenolic resins in decorative paints appears to have been fairly widespread throughout the 1930s. By 1930 they were well established in the United States and steadily gaining ground in the United Kingdom. Advertisements for quick-drying four-hour enamels appear with increasing regularity in the British trade press through the decade. The extent of their use in the United

Kingdom in particular is difficult to ascertain: some manufacturers claimed never to have used them due to their poor color,[2] whereas others stated that by 1930 all coatings in the United States and the United Kingdom that were advertised as quick drying contained them (Fonrobert et al. 1930: 90).

Claims made by British companies that they did not use phenolics must, however, be received with caution: as discussed in chapter 6, coatings based on synthetic resins were not well received in the United Kingdom, which made manufacturers reluctant to admit that their coatings contained them. But by the 1930s many U.K. companies were manufacturing quick-drying four-hour enamels, which almost certainly comprised phenol-formaldehyde resins (Boxall 1978b; Elliot 1981: 13; Lilley 1951: 307) (see plate 16). Walpamur's early Rockfast enamel paints were certainly based on phenolic resins, as was the Goodlass Wall Company's Valspar.[3] If Rohm & Haas's claim that by 1928 Amberol had displaced at least 90% of all other resins used in the varnish and enamel trade is indeed true, their use in the United States was widespread also (Hochheiser 1986: 44).

The use of decorative paints based on phenolics was restricted for the most part to the 1930s. Although largely replaced by alkyds in the postwar years, it is evident that phenolics were still being marketed for decorative applications as late as the mid-1950s: one 1953 author described rosin-modified phenolics, cresol formaldehyde modified with dehydrated castor oil, and esterified modified cresylic resin binders for use in decorative paints and four-hour enamels (Chatfield 1953: 782–97). And Rohm & Haas was certainly marketing its original Amberol resins, along with rosin-modified maleic types, for use in household decorative enamels in 1953 (Rohm & Haas 1953).

ALKYD RESINS

Alkyds are polyester resins that are co-esterified with oils or fatty acids to produce a range of baking and air-dry finishes for protective and decorative purposes. Alkyds are versatile resins that have been immensely successful since they became widely available in the years after World War II, and they remain an important binder to the present day. Although today's household paints based on alkyds are most commonly associated with gloss trim paints, they have also been used in the manufacture of flat and eggshell interior and exterior wall finishes in both solvent-borne and emulsified form. Alkyds have also been used as a modifying resin for styrene-butadiene, polyvinyl acetate, and nitrocellulose lacquers.

Manufacture

Alkyds are manufactured by means of a condensation reaction between a polybasic carboxylic acid and a polyhydric alcohol. For the purpose of decorative paints, the polyester resin is oil modified via co-esterification with fatty acids or their glycerides. Today an almost infinite number of acids and alcohols can be used in alkyd formulation and the structure modified to achieve a particular property, but household paints were and are most commonly made from phthalic anhydride (the acid) and pentaerythritol or glycerol (the alcohol).

The oil component confers important properties to the resin, controlling solubility, cross-linking potential, and compatibility with other resins. The fatty acid chains give flexibility to the otherwise brittle resin, both through their inherent flexibility, and also by reducing the functionality and therefore the capability of the molecule to cross-link and form hard insoluble polymers. Oil-modified alkyds can be divided into three main categories, distinguished by their oil length (proportion of oil to resin). The oil length is expressed as a percentage and denotes the number of grams of oil required to make 100g of resin. Alkyd resins can be broadly categorized into short (<45%), medium (>55%), and long (60%–80%) oil. The type of fatty acid or glyceride included also governs the properties of the resin: the oil might be of the oxidizing or nonoxidizing type, and the resins are thus referred to as drying, nondrying, or semidrying. Alkyds used in decorative paints are long-oil, oxidizing, drying resins.

Short-oil alkyds have high viscosity and require strong aromatic solvents to dissolve them. They are primarily used in baked coatings for applications such as Venetian blinds, metal cabinets, cans, caps, and autos. Medium-oil alkyds have lower viscosity and are soluble in aliphatic solvents; they are used in air-drying pigmented enamels and can be applied by roller or spray for maintenance, metal decorating, auto and truck refinishing, and sign painting. Long-oil alkyds are used in decorative enamels. Their viscosity is even lower than short- or medium-oil alkyds, and they are well adapted for brushing. They were and still are used for architectural finishes, flexible coatings (e.g., toothpaste tubes), and trim and trellis painting. Alkyds formulated from nondrying oils such as raw castor or coconut were also used for high-quality stoving finishes for white goods (Beavers 1949).

Alkyds are manufactured by either the fusion or solvent process. The fusion method is the oldest and involves heating the reactants together in a heated kettle. Early alkyds were manufactured in open kettles, which resulted in considerable loss of the reactants through volatilization and in resins that were poorly combined chemically, unstable, and dark colored. The introduction of closed kettles in the 1930s improved matters considerably, although some of the material was still lost. The fusion method is still used today for long-oil alkyds, although it is not recommended if a narrow specification for the resin is required (Holmberg 1987: 78–79). A much-improved resin can be manufactured by the solvent process, which was introduced in the early 1950s. Although this is similar to the fusion method, the reaction takes place in a closed kettle and in the presence of a solvent such as xylol, which distills and removes the water produced by the condensation reaction. The water then condenses and is removed, while the solvent is returned to the mixture where it continues its azeotropic action. The solvent method of manufacture minimizes loss of the reactants, prevents localized overheating, and results in a resin that is much better combined chemically, with superior color.

Oil-modified alkyds can be prepared using the triglyceride (the whole oil) or the fatty acids derived from it: the procedures are referred to as the monoglyceride process and fatty acid process, respectively. The monoglyceride process involves using triglycerides such as linseed, soybean, or safflower and heating them together with the polyol and dibasic acid. However, because the triglycerides react preferentially with the acids, it is necessary to carry out a controlled esterification process between the triglycerides and either the polyol or the acid before the condensation step, to ensure that the fatty acids are distributed evenly along the polyester backbone. The triglycerides can be reacted with either the polyol (alcoholysis process) or the acid (acidolysis process). In the fatty acid process fatty acids are used in the place of triglycerides. This process offers greater flexibility in choice of polyols and is more reproducible. The choice of triglyceride or fatty acid depends on the manufacturing process used, the acid and alcohol chosen, and the cost and availability of the oils or fatty acids.

Before World War II linseed was used to the exclusion of other oils in the manufacture of alkyds for decorative purposes. However, linseed discolors as it ages, making it unsuitable for the manufacture of pale-colored paints. This fault encouraged manufacturers to look for alternative oils, and in the years after the war oils such as soybean, safflower, oiticica, dehydrated castor, and tall oil fatty acids were used in the formulation of alkyd resins for a wide range of applications. The use of linseed oil thus declined steeply after the war, and by the mid-1950s most decorative alkyds that had comprised linseed were reformulated with soybean.

History

The first polymeric ester was discovered accidentally by the Swedish chemist J. J. Berzelius in 1847, when he heated glycerol with tartaric acid. Shortly after, J. N. van Bemmelen and A. V. Lourenco, among others, studied the reaction between glycerol and polybasic acids such as succinic and citric (Langton 1943: 16). In 1901 W. J. Smith prepared glyceryl phthalate by a condensation reaction between glycerol and phthalic anhydride, but its insolubility in any solvents offered no practical use in coatings (Martens 1961: 2). These early alkyd resins were hard and brittle and found use only as adhesives and electrical insulators (Boxall 1978b: 21).

Between 1911 and 1915 scientists at General Electric conducted extensive research into the reaction between glycerol and phthalic anhydride, and a number of useful resins resulted (Kienle and Ferguson 1929).[4] In 1914 W. C. Arsem, E. S. Dawson, and K. B. Howell found that more flexible resins could be produced if part of the phthalic anhydride was replaced with an aliphatic dibasic or monobasic acid. They developed fatty acid–modified alkyds that incorporated butyric, oleic, and stearic oils into their structure (Elliot 1981: 14). Glycerol could also be modified with castor oil to produce high-temperature baking varnishes. However, although tough, adherent, varnishlike films could be produced, these were of little value: they not only lacked build but were difficult to pigment, and their long baking time at high temperatures meant that they found limited use only for some electrical purposes; this, coupled with the expense of phthalic anhydride, made them uneconomical (Kienle 1949). However, in 1916 H. D. Gibbs et al. synthesized phthalic anhydride, and the abundant commercial supply certainly inaugurated the enormous amount of research undertaken into the development of alkyd resins at that time.

Following the discovery that part of the phthalic anhydride could be replaced by a fatty acid, further work with unsaturated oils, such as those of vegetable origin, was carried out, primarily by M. J. Callahan, van Bemmelen, and the Ellis-Foster Company in the United States (Bevan and Robinson 1951: 40). Around 1917 extensive research was being conducted into the bodying and drying of linseed and other drying oils, and although alkyd resins were insoluble in these oils, it was found that by introducing drying oil fatty acids to the reaction at the same time as glycerol and phthalic anhydride, a homogenous resin could be produced that would convert to an insoluble state on drying. These resins were soluble in low-cost aliphatic and aromatic hydrocarbon solvents, and the resulting films had good flexibility, adhesion, and durability (Martens 1961: 4).

The first application of alkyds in coatings was as plasticizing resins for nitrocellulose lacquers, which by the mid-1920s were rapidly replacing traditional oleoresinous automobile finishes. Nitrocellulose lacquers required the addition of a resin to ensure good adhesion to metal. Initially, natural resins such as dewaxed dammar were employed, but alkyds and rosin-modified alkyds were far superior to these, as they were not only compatible with nitrocellulose in all proportions but also imparted greater durability, gloss, and build. Alkyds were also an important addition to the nitrocellulose brushing lacquers described in chapter 4: DuPont's "Brush" Duco, which was introduced in 1926, was rendered brushable by the addition of an alkyd resin.

DuPont was at the forefront of research into the development of coatings based on oil-modified alkyds. This was largely due to the appointment of Callahan in 1924, who had worked on them at his previous employment with General Electric. DuPont's first attempt to commercialize alkyd resins was in 1925, and by 1926 it had introduced its first Dux primers based on an oil-modified alkyd; these replaced the slow-drying oleoresinous undercoats that had been used under nitrocellulose Duco finishes for automobiles (Zintl 1947: 76).

DuPont began its research into decorative paints based on alkyds in March 1927. The work was conducted by Gordon D. Patterson, a research chemist.[5] His initial work compared the performance of alkyds modified with tung and/or linseed oil fatty acids and ester gum modified with tung oil: although the alkyds were undoubtedly superior, they had many shortcomings, including acidity, poor color, yellowing, poor brushing, dragging, poor stability in the can, heavy skinning, granulation, and high cost. Indeed, the expense of these early alkyds prompted DuPont to temporarily halt work on their development in February 1928, the stated reason being that research indicated that there was "no market for a paint costing more than existing products, irrespective of the outstanding properties to be offered."[6]

The cost of fatty acids compared to triglyceride prompted DuPont to formulate a resin in which 50% of the fatty acids were replaced with refined oil. The incorporation of whole oil was allowed by the increased use of the alcoholysis method of manufacture.

Despite great strides being made, by the end of 1928 alkyds were not ready for introduction to the market. Although they were undoubtedly durable and resistant to insects, weather, and dirt, they were difficult to use at high temperatures due to fast solvent evaporation, had low spreading rates per gallon because of their poor hiding power, and were more expensive than traditional oleoresinous binders. Despite these shortcomings, the first Dulux oil-modified alkyd paint was offered to the industrial trade in the summer of 1929. Although the durability of the new binder was not in question, these initial paints were unsuccessful due to both poor brushing qualities and the fact that they were pigmented with the inappropriate lithopone:[7] when exposed to sunlight, lithopone darkens from the reduction of zinc sulfide to zinc metal.

Throughout the summer of 1929 Patterson and his colleagues worked on the pigmentation of Dulux. They found that a combination of titanium dioxide, which had been incorporated into paints since about 1927 (Snelling 1952b: 59), with a small addition of lithopone or zinc oxide was the most promising for white paint. Once the pigmentation of Dulux had been improved, the paint was formulated for industrial purposes, including signs, refrigerators, car wheels, railway passenger and freight cars, railway equipment, gasoline pumps, and service stations. But although durable, these oil-modified alkyds were not popular with the trade, as they still suffered from insufficient hiding power (DuPont 1949: 2). Although exterior trim paints were formulated at this time, they were found to be unsatisfactory and were later withdrawn from the market (Zintl 1947: 78).

These early alkyds suffered serious defects: they were unstable and dark colored, had poor water resistance, behaved disappointingly with a number of commonly used pigments, and were poorly combined chemically. This resulted in a host of problems, including high acid numbers, rapid increase in the viscosity of the resin solution, and, occasionally, actual separation of the resin from the solvent (Troutman 1943: 342). These defects were caused by several factors: in addition to the inappropriate pigmenting of early paints, they were manufactured using the old-fashioned open pot methods of the traditional varnish manufacturers. Open pots had poor temperature control and allowed evaporation of the reactants during manufacture. This was entirely unsuited to the manufacture of alkyds, which require tightly controlled conditions to produce high-quality, properly combined resins. The introduction of closed kettles in the early 1930s helped to control temperature and minimize evaporation and thus improved some of these problems.

In 1930 Dulux refinishing enamel was introduced for automobiles in the United States, which immediately became popular with car dealers. Although slower drying than Duco, it was considerably cheaper and therefore popular for repainting lower-priced cars that did not warrant a Duco finish (DuPont 1949: 2). In May 1931 the first exterior house paint, Dulux GA-11, was released by DuPont. It comprised a 62% linseed-modified alkyd resin, pigmented with Titanox B (titanium dioxide); the tung oil component had been dropped, as it would be from most oil-modified alkyds for decorative purposes. The sales department began to promote Dulux exterior paint that year, expounding its excellent durability. Yet problems with the product continued. The alkyd component of the resin comprised glycerol and phthalic anhydride, but glycerol produces a hard film that is impervious to moisture. If moisture is present behind the paint film, as can happen in poorly constructed houses, instead of diffusing through the coating, the moisture is held beneath, causing the paint to blister and eventually flake off in large sheets. It is clear that in its rush to be the first to launch this product on the market, DuPont had not adequately tested Dulux, and the company was obliged to foot the bill for repainting the faulty houses, at a cost of approximately $250,000 (Zintl 1947: 91).

Throughout 1931 Patterson and his colleagues worked on improving the pigmentation of alkyd resins. The first titanium dioxide pigments were available in anatase form only, which chalks badly. Although at that time controlled chalking was considered desirable in white paints, excessive chalking resulted in early breakdown of the coating and a white powdery deposit on the surface of colored paints. The introduction of the rutile form of titanium dioxide improved this fault significantly, as did the use of pigments coated with water-repellent substances such as soaps of cadmium, barium, zinc, and aluminum or oil-soluble resin or wax. By 1934 these improvements had been applied, and DuPont introduced Dulux Super White, a new finish for home interiors. One of the first advertisements appeared in *American Painter and Decorator* in March 1934, expounding the paint's nonyellowing properties, quick drying, and easy application. Once Dulux house paints had been successfully formulated and marketed, the binder of DuPont's "Brush" Duco, which had been unsuccessful on a nitrocellulose base and currently contained an alternative, quick-drying medium, was replaced with the new Dulux alkyd resin (see plate 17). In 1937, after several years of field testing, improved Outside White for exterior use was released. At the same time, Dulux Super White was improved for interior use; there is not, however, any indication as to how the interior and exterior Dulux differed.

Throughout the 1920s and 1930s, while DuPont was developing Dulux resins, a long-standing agreement to share technical information between itself and the U.K. company ICI ensured that new patents and processes were introduced swiftly to the British market via NCF. NCF was a joint company established in 1926 by DuPont and ICI specifically for the manufacture of nitrocellulose Belco finishes, the British equivalent of Duco. Therefore, Dux industrial undercoats based on alkyds were launched in Britain in 1929; the Dulux trade name was registered by NCF on November 27, 1930; Dulux industrial coats to replace air-drying and baking enamels were introduced in May 1931 (*Synthetic and Applied*

Finishes, March 1931, 171, 23); Dulux Glossy decorative paint was introduced in 1932 ("Story of Paints Division" n.d.: 9); and in the same year Dulux eggshell and flat were introduced as interior wall finishes ("Two New 'Dulux' Products" 1932: 70).

ICI's Dulux paint underwent several changes in formulation and pigmentation in the 1930s, and although documentation in the ICI archive does not clarify what these involved (Snow n.d.), it presumably followed a course similar to the improvements made by DuPont. Dulux Glossy comprised phthalic anhydride, glycerol, and linseed fatty acids and had an oil length of 60% (Lilley 1959: 1). Although the durability of Dulux was not in question, the paint suffered defects similar to those reported by DuPont at that time: a thin consistency that tended to sag on vertical surfaces, poor brushability, and, due to the high content of volatile thinners, rapid setting time in warm conditions (Bevan and Robinson 1951: 358). Despite their excellent durability, the defects of these prewar alkyds made them unpopular with industry, and ICI experienced some difficulty selling its new product to this section of the market ("A Short History" 1979: 530).

The agreement between DuPont and ICI undoubtedly ensured the early introduction of alkyd-based house paints in the United Kingdom. The agreement continued until 1948, when it was severed by the enforcement of the Sherman Anti-Trust Act. The act ordered the dissolution of companies jointly owned by ICI and DuPont, and the exchange of technical information ceased.[8] The agreement had been remarkably one-sided in terms of information exchange, however, and ICI conceded that it had contributed nothing to the development of oil-modified alkyds before 1939, relying instead entirely on DuPont technology ("Story of Paints Division" n.d.: 11). Indeed, it would seem that ICI's approach to research and development in coatings was haphazard to say the least during these years, and ICI reported that as late as the 1940s, when research and development were being conducted in earnest, the "work was driven along without much pre-planning and paper work, and without continual reporting" (Lilley 1959: 6).

Development of Alkyds during World War II and Beyond

The onset of World War II in 1939 once more disrupted and altered the direction taken by paint manufacturers. Companies in Britain and America were put on a war footing once more, and both ICI and DuPont were called on to produce large amounts of explosives and coatings for the war effort. American companies continued to produce coatings based on synthetic resins such as phenol-formaldehyde and alkyds, but these were restricted to military use only. Production for the civilian market all but ceased, and by 1943 it was reported that 75% of paint produced in the United States was for military purposes (Burr Price 1943: 81). For ICI, the production of alkyd-based Dulux decorative finishes ceased completely during the war years, not only due to a shortage of raw materials but also because the paint department was required to manufacture vast amounts of the camouflage and gas-resistant coatings required by war ("Story of Paints Division" n.d.: 12). Throughout the war decorative paints in both countries—if they were manufactured at all—were formulated exclusively from naturally occurring oils, gums, and resins or their derivatives, and it was not until the late 1940s that oil-modified alkyds were reintroduced.

The research and development of postwar alkyd resins was again driven by cost and availability of raw materials. At the cessation of war in 1945, ICI found itself left once more with plants capable of producing large quantities of coatings and explosives but no large-scale market in peacetime. In their capacities as explosives manufacturers, both ICI and DuPont had been producing considerable quantities of the alcohol pentaerythritol, which had become commercially available between 1935 and 1940 (Spiller 1952) but was too expensive to be considered as an ingredient for alkyds in the prewar years. Nobel's development of a cheaper method of manufacture almost certainly played a vital role in inaugurating the research into its development as a coatings constituent by ICI (Snow n.d.). ICI's research into the development of alkyds resumed in 1945, using pentaerythritol in place of glycerol. Small batches of a pentaerythritol alkyd were definitely made by ICI in 1945 and again in 1946, with a large batch being made in 1948, suggesting that experiments had been carried out on the earlier batches before large-scale manufacture commenced (Lilley 1959: 3–4).

Although both alcohols have advantages and disadvantages, there are three main benefits to using pentaerythritol over glycerol. First, pentaerythritol has a functionality of four as opposed to glycerol's three, thereby allowing greater amounts of oil to be incorporated into the resin. This not only imparts greater flexibility to the film and improves the resin's rheological properties but also makes manufacture more economical, as oil is less expensive than either glycerol or pentaerythritol. Second, their water resistance is far superior to that of glycerol alkyds; and third, they have better gloss and gloss retention. Although the use of pentaerythritol for alkyd manufacture had been patented early in the twentieth century, its expense at that time had discouraged rigorous investigation. There is evidence to suggest that both the United States and Germany made use of pentaerythritol alkyds during World War II, but it appears that this was only because glycerol was in short supply, and no technical advantage was perceived or sought.

ICI initially encountered problems with the use of pentaerythritol, most of which were caused by the scientists' lack of understanding of the chemical reactions taking place; pentaerythritol could not simply be substituted for glycerol on a weight-for-weight basis, which had governed the first attempts at its use. Furthermore, it is apparent that the formulation of these early alkyds was restricted by the capability of the plant. During the 1940s ICI made alkyds using the fusion process and by the fatty acid method, which required a long cycle. Both these factors were reportedly unsuited to the available plant (Lilley 1959: 5), which in itself must have affected the quality of the resin; in 1948 pentaerythritol alkyds suffered from bloom, floating, flooding, and unsatisfactory drying.

However, ICI reintroduced decorative Dulux in December 1949, based on a phthalic anhydride, pentaerythritol, and linseed oil. A report on the 1949 United Kingdom Building Exhibition welcomed its return after a ten-year absence and stated that the properties were superior to the prewar version; the paint was available in gloss finish, interior enamel, and eggshell finish ("The Building Exhibition" 1949: 93). But this time numerous other British companies had become involved in the development of oil-modified alkyds. By the late 1940s the processes required to make them had been widely patented, and those without adequate research and development facilities could simply buy a license to manufacture a resin developed by another company. Alternatively, the premixed resin could be bought from any number of chemical companies and blended by the paint company.

On the introduction of pentaerythritol alkyds in the late 1940s, the use of fatty acids for alkyd manufacture once again increased, as pentaerythritol does not allow the use of refined oils that contain glycerine (Earhart 1954: 82). The use of fatty acids had many other advantages over whole oils, including flexibility in formulation and the end properties it allowed, a reduced kettle time (Schroeder 1952), and a suitability for the solvent method of manufacture that became widespread in the early 1950s.

The availability of many different types of oils in the postwar years also significantly affected the development of alkyd resins. Before the war linseed was used to the exclusion of other oils. However, although an excellent drier, linseed oil discolors on aging and is unsatisfactory for pale-colored paints. This fault encouraged companies to look for alternatives, and as oil-processing technology developed, so too did the range of semi-and nondrying oils that were formulated into alkyds, which by then included safflower, oiticica, dehydrated castor, and tall oil fatty acids. The use of linseed oil therefore declined steeply after the war, and by the mid-1950s most alkyds that had comprised linseed were reformulated with soybean.

Examination of the DuPont archives reveals that the company experimented with soybean triglyceride in alkyd formulations from the mid-1930s (Coolidge 1936), and soybean fatty acids were first developed and used in the United States during World War II. Due to its expense, soybean was seldom found in British paints before the late 1950s, and then it was used only for pale-colored paints; dark tints and cheaper paints for the export market continued to be manufactured with the considerably cheaper linseed oil.[9] But by the early 1960s soybean was the most widely used oil in the manufacture of alkyds in both Britain and the United States.

By the early 1950s numerous companies in both countries were producing high-quality oil-modified alkyd resins. Having lost their competitive edge by no longer being alone in their manufacture, both ICI and DuPont found their products superseded by their rivals', which were often not only of better quality but cheaper as well. This trend was partly the fault of the companies: DuPont reported that in the mid-1950s the technical support of trade sales had undergone drastic cutbacks in previous years, with the number of chemists on the assignment being reduced from seventeen in 1951 to just eight in 1955. The results of these changes became manifest as poor sales; DuPont's Outside White House Paint and white enamels, once considered the standards of the industry, were overtaken by competitors manufacturing better products at lower prices. It is reported that a concerted effort on behalf of DuPont reversed this trend by the late 1950s (Zintl 1947: 7). ICI's sales also began to suffer when other companies began to manufacture alkyd resins, which were similarly often of better quality: both companies had suffered the typical setbacks of being the first to introduce to the market a new type of product.

The most rapid advances in the development of oil-modified alkyds undoubtedly occurred in the years immediately after the war not only through the use of pentaerythritol, better pigmentation, and the incorporation of pale, semidrying oils but also due to improvements in manufacturing techniques. By the early 1950s the fusion technique, and all its inherent problems, had largely given way to the solvent process of manufacture, which resulted in a paler, higher-quality, and generally much improved product.

Properties

Oil-modified alkyds have many good properties that have ensured their continuing popularity, including fast drying, good color and color retention, and excellent gloss, build, and durability. Decorative alkyds dry by oxidative polymerization in much the same manner as the oil or oleoresinous binders described in chapter 2. However, their high molecular weight, high functionality, and large molecule size means that they require far fewer cross-links than traditional oil-binding mediums to harden, resulting in a considerably faster curing time (alkyds are typically touch dry in a matter of hours). These are also the properties that enable alkyds to dry when manufactured from semidrying oils such as soybean.

Alkyds are cheap to manufacture and soluble in inexpensive solvents, and the resin itself is low in odor (although the solvents required to dissolve the resin have odor). They are easy to apply by a number of methods, including brush, roller, dip, or spray. They have excellent pigment wetting and dispersing properties and are compatible with a wide variety of other compounds, making their capacity for modification almost unlimited (Martens 1961: 12). They also have good sealing properties and better flow than alternative systems, and when used in solvent-borne systems, they are resistant to freezing, have good package stability, and retain their viscosity well—important considerations for retailers. They are washable within twelve to eighteen hours and can be scrubbed within forty-eight hours ("Cook County" 1954: 74–75). Alkyds can be used in air-drying or baking finishes and have many industrial and domestic applications.

The properties of the final resin depend on the constituents used. Therefore, in much the same manner as oleoresinous varnishes, short-oil types have better gloss retention, and long-oil types exhibit better weathering. And whereas long-oil alkyds have properties similar to the oil employed, short-oil alkyds have the characteristics of the polyester resin. If unsaturated oils are used, the coating dries quickly but has poor weather resistance and yellows on aging. Lower-oil-content alkyds have poorer drying properties but better stability, gloss retention, and weathering (Stoye, Freitag, and Beuschel 1996: 66). The acid and alcohol used also have a significant impact on the final properties. Alkyds formulated with glycerol are harder than pentaerythritol alkyds and are thus suitable for applications such as undercoats, which must withstand sanding. Although pentaerythritol alkyds are faster drying and more resistant to alkalis, glycerol alkyds have better stability, adhesion, and compatibility with other film-forming materials (Beavers 1949: 42).

Alkyds have good resistance to humidity and intermittent water exposure, but they are not suitable for immersion in water unless modified with certain compounds such as phenolic resins. Alkyds have fair resistance to petroleum solvents and oils, but films are softened by alcohols and aromatic hydrocarbons and removed by ketones, esters, and chlorinated solvents. They are also attacked by concentrated acids, oxidizing agents, and strong organic acids. Although alkyds give good protection against neutral salts, they have poor alkali resistance (Martens 1961: 67). Susceptibility to hydrolysis is indeed the biggest flaw of alkyds, a fault that is inherent in the chemical structure of the alkyd, as all polyester resins contain an ester linkage that is susceptible to decomposition by water, especially in the presence of acids and alkalis. It was in an effort to overcome this defect that modified alkyds were developed in the 1940s. Modifying compounds included styrene, silicone, vinyl toluene, urethane, and methyl methacrylate and other acrylic acids.

Although widely manufactured, there is little evidence to suggest that styrenated alkyds were used in decorative paints, although they were reportedly used for low-cost fast-drying enamels, baking coats, and undercoats (Haines, Griffin, and Da Valle 1954). Urethane-modified alkyds were certainly used in decorative paint manufacture. Reaction with isocyanate confers rapid drying time, hardness, good elasticity, excellent abrasion resistance, high gloss, and much-improved water and chemical resistance, including in alkaline conditions. Urethane-modified alkyds were and still are used for interior and exterior wood coatings and also for marine and decorative paints (Stoye, Freitag, and Beuschel 1996: 74). Alkyds have formed the basis of a wide range of household paints, including gloss trim paints, kitchen and bathroom paints, flat and eggshell wall paints, and exterior paint for protecting wooden houses.

Flat Wall Paints Based on Alkyds

Today decorative paints based on alkyds are associated almost exclusively with gloss finishes for interior and exterior use. However, soon after their introduction they were also used for interior and exterior flat paints. Both emulsified and solvent-borne alkyds were used in the formulation of durable, flat wall paints from the late 1930s, although by the early 1950s they had largely been superseded by synthetic polymer emulsion paints for this purpose.

Emulsified Alkyds

Emulsified alkyd resins were first introduced just before World War II in Europe (Martin 1947: 148) and became commercially available in the

United States soon thereafter. There has been a resurgence of interest in them recently, largely due to increasing restrictions on VOC emissions. The first alkyd emulsions were used as a binder in their own right for interior and exterior paints, and small quantities were sometimes added to emulsion paints based on styrene-butadiene and PVAc in the belief that they increased durability. Emulsified alkyds were useful where an odorless, fast-drying, yet durable and washable finish was required. However, they are seldom mentioned in paint trade literature, and the extent of their use for household paints is unclear.

The limited use of alkyd emulsions must have been caused in part by the difficulty of their manufacture: they are delicate systems that are complex to formulate, and it is evident that they did and indeed still do provide a real challenge for manufacturers. For the first manufacturers, many of the problems arose from the slightly alkaline conditions required to achieve a stable, finely dispersed emulsion. This is problematic for two reasons: first, alkyds have a tendency to saponify and form water-soluble salts in basic conditions, which meant that only the least reactive, long-oil type of alkyd could be emulsified successfully; and second, the most effective emulsifier, casein, degrades in water and basic conditions, causing a reduction in the viscosity of the emulsion. Although cellulose ethers and alginates were alternative, more stable emulsifiers, they were nowhere near as effective. Moreover, the mere presence of emulsifier was detrimental to the film's properties, affecting water resistance, color, durability, and permeability. In order to minimize its effect on the water resistance of the film, the emulsifier had to be a low enough quantity that it could be dissolved in the resin phase as the film formed. Indeed, all the additives required for such systems were to some degree hygroscopic and had the capacity to seriously compromise the water resistance of the dried film.[10]

The properties of alkyd emulsions varied enormously, depending on how they had been formulated, but they were usually supplied at 40% to 50% solids, and only drying oil-modified alkyd resins that had good durability and color retention were recommended for emulsifying; the dry film would then have similar properties to a solvent-borne film of the same composition. Many alkyd emulsions were supplied in a heavy paste form, to be thinned with water just prior to use. Preservatives were added during manufacture to prevent putrefaction and mold growth. Despite their shortcomings, emulsified alkyds had many good properties when compared to other systems: they were easier to apply and gave cleaner shades than oil-based paints, and they were more durable and scrubbable than alternative aqueous systems; they could also be applied to damp and porous surfaces. Alkyd emulsions dried via a two-step process that involved evaporation of the water phase, followed by coalescence and oxidation of the resin phase. The film had initially to remain permeable to moisture to allow the water to evaporate, which was followed by oxidation and hardening of the alkyd. However, hardening was slower than for equivalent solvent-borne systems. As with all alkyd systems, the coatings were sensitive to alkaline hydrolysis.

The extent of use of alkyd emulsions in decorative paints is difficult to ascertain. In the United Kingdom certainly this was partly due to the reticence of British manufacturers to advertise the components of their paints, but they certainly found use for painting office blocks, hotels, schools, and hospitals, which required a low-odor coating that was durable and quick drying. However, alkyd emulsions do not appear to have been widely used and were superseded by the vinyl emulsion paints that were introduced in the late 1940s (see chapter 6).

Solvent-Borne Alkyd Wall Paints

Solvent-borne wall paints based on alkyds were first available in the early 1930s but did not gain popularity until the early to mid-1950s, after formulation problems had been resolved. By the mid-1950s they were in direct and fierce competition with emulsion paints based on styrene-butadiene and PVAc. Alkyds had many advantages over emulsion paints from the perspective of both the user and the manufacturer: they had better hiding power than equivalent finishes and therefore could cover in one coat; they were quick drying, immediately scrubbable, and very durable; and they also offered an almost unlimited choice of pigments, freedom from freezing, and compatibility with oil colors and enamels. Alkyds were also easier to apply than oil paints, although inferior to emulsion paints in this respect (Bragdon 1954: 86). Furthermore, it was possible to produce a range of paints—from gloss trim through to flat wall—from a single binder that was available in matte or gloss, and could be intermixed to achieve any finish in between (*American Paint Journal,* July 14, 1952, 52–53). By the early 1950s many companies were thus producing alkyd binders for a range of interior paints. By 1960, well after alternative waterborne systems were well established, there was reportedly a "hard core" of professional painters who continued to use alkyd paints (Williams 1960: 158).

However, despite their many good properties, alkyd paints had a significant drawback that limited their appeal to the home user in particular: the hydrocarbon solvents they were dissolved in not only made cleanup an unpleasant task, but their strong odor permeated the house, so that interior painting could only be carried out in fine weather, when it was possible to ventilate the house. This was an especially pertinent point, as the emulsion paints with which they were in direct competition were easy to clean up with water and low in odor. Therefore, in an effort to overcome the problem of odor, manufacturers developed odorless alkyd paints.

Odorless alkyd binders appear to have been restricted to the U.S. market, as there is no discussion of them in British trade literature. This should not come as a surprise, as at that time few British manufacturers had attempted to exploit the retail market. Odorless alkyds were introduced to the U.S. market around 1952, and from that time numerous companies began to advertise odorless flat, gloss enamels, and eggshell finishes. Although the idea was a good one, odorless alkyds suffered from many drawbacks that rendered them obsolete fairly rapidly: they were expensive to produce, and manufacturers required special equipment, all of which had to be kept spotlessly clean ("Odorless Solvents" 1952: 30). Deodorized solvents not only had very low solvency and poor compatibility but also a tendency to regain their odor and deteriorate in the can. Although odorless driers were available, in many instances it was only the solvents that were odorless; both the driers and the resin carried an odor. Moreover, if the user wished to thin his paints, he had to do so with expensive odorless solvents, and cleanup up could not be done with water, as it could with emulsion paints. Finally, although the paint in the can might be odorless, alkyds emit an odor during oxidation and drying ("Husen" 1952: 9–11).

That odorless paints were geared specifically to the home user is evident from the advertising that accompanied their introduction: one manufacturer redesigned its packaging specifically to appeal to women and family users (*American Paint Journal,* October 13, 1952, 96–97); another showed a woman sniffing the paint she had just applied to the wall (*American Paint Journal,* July 7, 1952, 74–75). The latter ad in particular was perhaps disingenuous, as despite manufacturers' claims that painting could be done with the windows closed, the National Paint, Varnish, and Lacquer Association advised "not to take the chance" and to paint with good ventilation (Schultz 1969: 198). It should perhaps come as no surprise that by the mid-1950s these products had been all but superseded by emulsion paints.

Thixotropic Alkyds

Thixotropy was an important development in paint production. It was conferred by reacting the resin with polyamide. Thixotropic paints have a thick, gel-like consistency, which improves their brushing properties and prevents sagging on vertical surfaces. It also enables the paint to be used directly from the can without stirring and has the added benefit of preventing pigment from settling during storage. Thixotropic paints were reportedly first introduced to the U.S. market in 1948, but they were not manufactured on a large scale until 1955 ("'Gelled' Paint" 1955: 69). They were initially formulated for interior use only and proved immediately popular with the home user. One of the first manufacturers in the United States reported that demand had increased each year since 1948, until by 1955 these paints represented 45% of the firm's total sales. The first thixotropic alkyds were reportedly based on a binder called Burnok, which comprised a polyamide resin made from soybean oil. Thixotropic paints became available in the United Kingdom in 1956, at which time a number of manufacturers introduced versions to the U.K. market.

Availability and Use of Alkyd Resins

The excellent properties of alkyd resins were recognized immediately on their introduction in the mid- to late 1920s, and industrial coatings gained considerable success in the early years. Before World War II coatings comprising alkyds modified with compounds such as urea-formaldehyde were used widely for coating autos, refrigerators, and similar products ("Alkyd Resins" 1946: 52). Decorative paints based on alkyds did not enjoy much success in the early years: although they were undoubtedly durable, they had many application and stability problems and were more expensive than traditional oleoresinous coatings. Although the market for them was initially small, as the paint was improved, sales gradually increased: between 1933 and 1938 demand for DuPont's Dulux White rose from one million to six million gallons per year (Zintl 1947: 80), and by 1937 ICI's sales of Dulux were about eighty thousand gallons per year, with one to two hundred thousand gallons exported to British territories; these finishes were available in matt, eggshell, and gloss ("Story of Paints Division" n.d.: 11).

Few British companies manufactured oil-modified alkyds in the 1930s: Jenson & Nicholson introduced alkyd-based Robbialac paints in the late 1930s,[11] and British Paints conducted research into alkyd resins in the 1930s, but their production was certainly limited in both Britain and the United States. Prewar decorative finishes in both countries comprised glycerol, linseed oil fatty acids, and phthalic anhydride.

When war broke out in 1939, the production of alkyds for the decorative markets ceased completely in both countries, although they were still manufactured for military use ("Alkyd Resins" 1946: 52). It is evident that the shortage of raw materials was acute in the years after World War II, and in the United Kingdom, Dulux finishes were not reintroduced until December 1949. However, ICI continued to advertise Dulux in the intervening years, with copy such as "the supply position is improving, so please write for news of 'Dulux' finishes" (*The Decorator,* April 15, 1946, 47).

Decorative paints based on alkyds became widely available once more in the mid- to late 1940s, and most companies waited until that time to enter the oil-modified alkyd field: it is thought that Berger did not put its Pompeian range of paints onto an alkyd binder until after World War II.[12] In the United States, Benjamin Moore's Impervo Enamel was put on an alkyd binder in about February 1946,[13] and Sherwin-Williams introduced its first oil-modified alkyd, Kem-Glo, in 1949 (McDermott and Dyer 1991: 63).

Glycerol was still the most widely used alcohol for alkyd manufacture in the early 1950s, as it had a number of advantages: it occurred naturally in the oil and could be used directly in the alcoholysis process of manufacture; it was more economical than alternatives, as it contains the highest amount of active hydroxyl per weight; it allowed flexibility in formulation, as it can be used to formulate very long to very short alkyds; and it had been in use for the longest time so there was a good understanding of its properties and performance (*American Paint Journal,* October 27, 1952, 38–40). However, the recurring shortages of glycerine encouraged manufacturers to develop other types of polyhydric alcohols, of which pentaerythritol was undoubtedly the most important in decorative paint applications. By the early 1950s pentaerythritol was becoming increasingly popular for formulating decorative paints. By that time alkyd manufacturing techniques had also been perfected, which, in combination with the use of pale, semidrying oils, resulted in a vastly improved product, and sales of alkyds increased dramatically.

As was the case with coatings based on other synthetic resins, in the early years alkyds appear to have been more widely used and readily accepted in the United States than in Britain. The reasons for this are complex, but it was due in part to the much larger retail market in the United States, whose paint companies had been gearing their products to the home user for many years; it was also due to the fact that U.S. consumers generally seemed more willing to accept coatings based on synthetic resins. The sale of alkyds was certainly stifled by the notoriously conservative decorating industry in Britain, to the extent that many companies did not advertise what was in their paints at all. For example, none of the 1930s British ads for Dulux stated what the binder was: in 1931 it was described simply as "new" (*Synthetic and Applied Finishes,* March 1931, ii) and by 1934 as "synthetic" (*The Decorator,* May 15, 1934, 53). It was not until the early 1950s that any companies advertised the fact that their paints were based on alkyd resins (*The Decorator,* November 15, 1951, 98). The acceptance of synthetic resins in the United States and the United Kingdom is discussed in greater detail in chapter 6.

By the 1950s decorative paints based on alkyd resins were in direct competition with traditional oleoresinous gloss paints and emulsions and enamels for walls. Their success as a binder for flat wall paints was certainly curtailed by the popularity of emulsion paints, which were low odor, easy to apply, and easy to clean up. However, alkyds were used when good durability was required and therefore remained popular for applications such as kitchen and bathroom paint. They were also preferred by many decorators, for whom low odor and ease of cleanup was not of primary importance. However, the most successful application for decorative paints based on alkyds was as a replacement for the natural gums and resins used in varnish and gloss paint manufacture. And although alkyds did not completely replace traditional oleoresinous finishes until well into the 1960s, by 1954 they had outsold naturally occurring resins in the United States for the first time ("The Markets" 1954: 50), a trend that continued throughout the next decade. By the 1960s traditional oleoresinous finishes were no longer economically viable, and they were virtually completely replaced with alkyds in this field. By 1961 alkyds dominated the U.S. coatings industry, and over 50% of all resins used in surface coatings manufacture comprised them (Martens 1961: 6–7). They remain one of the most important paint binders to the present day.

Chapter 6

Emulsion Paints Based on Synthetic Resins

SYNTHETIC RESIN EMULSIONS

Waterborne paints have always been popular, as they are easy to apply, easy to clean up, and nontoxic. They also produce clear colors and pleasing soft, flat finishes that minimize wall defects. However, traditional water paints lacked water resistance and durability, and durable oil-based equivalents had an unpleasant odor, took time to dry, dragged on the brush, and had insufficient moisture permeability for many applications. Although the oil and alkyd emulsions that were introduced in the 1930s approached oil paint in terms of durability, they also took time to dry properly, were impermeable to moisture once dry, and, depending on the binder used, were susceptible to alkaline hydrolysis.

It was not until the introduction of emulsion paints based on synthetic polymers in the late 1940s that the good properties of both oil and water paints were combined in one binder. Emulsion paints based on synthetic polymers were easy to apply, easy to clean up, quick drying, durable, washable, moisture permeable, and nonsaponifiable, and they could be recoated many times: they proved immediately popular with the professional painter and the home user alike. Indeed, their introduction boosted the do-it-yourself market dramatically. Their primary use was and still is for interior flat wall paints, exterior masonry paints, and primer sealers. The three binders used in household paint manufacture are polyvinyl acetate, which was introduced in the late 1940s; styrene-butadiene copolymer, also introduced in the late 1940s but used in the United States alone; and acrylic, which was introduced in the mid-1950s; plasticized polystyrene binders were also available in the United States in the early 1950s. All are vinyl resins, in that they are formed from monomers that contain the functional group $CH = CH_2$.

History

An emulsion is a dispersion in which two immiscible phases are held in suspension. In the case of emulsion paints, a hydrophobic polymer is suspended in water. Suspension is achieved through the use of emulsifiers, surfactants that have a hydrophobic and hydrophilic portion to their structure. The hydrophobic portion is attracted to the polymer, whereas the hydrophilic portion is attracted to the water, enabling the polymer to be suspended in the water phase. The smaller the particle size, the greater the quantity of surfactant required to keep the polymer dispersed in the water. Emulsions also require various additives (see Emulsifiers, Surfactants, and Protective Colloids, below).

Although vinyl polymerization techniques were developed in the nineteenth century, the monomers did not become commercially available until the twentieth, and emulsion polymerization was not perfected until during and after World War II. Therefore, household emulsion paints based on synthetic polymers were not introduced until the late 1940s. A number of vinyl monomers were available for the manufacture of emulsion polymers, including butadiene, isoprene, acrylonitrile, styrene, vinyl toluene, vinyl chloride, vinyl acetate, vinylidene chloride, vinyl pyridine, and acrylates. These could be used alone or in combination and were chosen to impart specific properties to the paint. For example, vinyl acetate promoted good adhesion, acrylonitrile offered good grease resistance, and vinyl chloride had good aging properties (Fowler and Westerhoff 1952).

As is so often the case, the development of the different binders for household paints was driven purely by cost and availability of raw materials: styrene-butadiene was used in the United States alone, as the country's vast synthetic rubber industry provided a ready supply of cheap monomers; Germany developed PVAc resins in the 1930s as a result of acute wartime shortages of linseed oil; and Rohm & Haas's development of acrylic emulsions in the early 1950s was in part driven by the desire to find an outlet for acrylate monomer, thereby justifying the construction of an improved monomer plant. Because of the association with rubber, styrene-butadiene paints were called latex paints in the United States, a name that has since been used to describe all synthetic polymer emulsion paints there, regardless of polymer type. For clarity, the term *styrene-butadiene* is used throughout this text, and *emulsion* is used as the generic term to describe all waterborne synthetic polymers.

The initial development of emulsion paints was largely carried out by those already involved in water paint production, as it was they who understood the intricacies of water-based systems and had suitable plants. But such was the success of emulsion paints that all paint manufacturers quickly recognized their potential, and most entered the market fairly swiftly. However, emulsion paints require quite different formulation from oil-based paints, and companies wishing to manufacture them had to furnish themselves with an understanding of principles and ingredients they had never before encountered, including protective colloids, sequestering agents, antifoam agents, pH control, emulsion stability, and freeze-thaw stability. Emulsion paints also suffered from microbial attack, so they required preservatives, and their tendency to both foam and freeze meant that they required careful formulation, transportation, and storage; for these reasons they were initially not popular with either formulators or distributors (Scholl 1952c: 60–62). Despite these problems, their ease of application and inexpensiveness made the paints immediately popular with consumers.

As the emulsions discussed here share many common features, the general principles of emulsion polymerization are outlined below, before detailed discussions of the individual binders.

Overview of Emulsion Polymerization

Synthetic polymers are created by heating the constituent monomers in the presence of catalysts such as persulfates or oxides. As the resulting reaction is highly exothermic, a diluent is added to control it. The diluent can be either an organic solvent or water: a solvent results in solution polymers; water, in emulsion polymers. Although both types have been formulated into coatings, it is the emulsified polymer that finds most extensive use and forms the basis of waterborne household paints.

For successful emulsion polymerization, numerous additives are required: catalysts initiate the polymerization reaction, emulsifiers (nonionic or ionic surfactants, or a mixture of the two) control the molecular weight of the resin and stabilize the emulsion once it has formed, and protective colloids also help to stabilize the emulsion and confer good rheological properties to the paint. It should be noted that protective colloids are not required in all emulsion systems.

Emulsion polymers can be manufactured by the batch or the delayed addition process. The batch process is the oldest and simplest method and entails charging the monomers, emulsifiers, and protective colloid (if required) together in the kettle. However, as the resulting polymerization reaction is strongly exothermic, it is difficult to provide adequate heat transfer to remove the heat generated. Furthermore, the final emulsion has a relatively large particle size of 2–3 microns. A better product can be manufactured by using the delayed addition process, which consists of dissolving the emulsifiers and protective colloid in water, adding some of the monomer and initiators, heating, and then adding the rest of the monomer at various intervals as polymerization progresses (Warson 1972: 379). The delayed addition process results in resins with a considerably finer particle size and, if a copolymer is being created, also ensures that the different monomers are evenly distributed within the polymer molecule. Chain regulators, buffers to control the pH, and thickening agents to increase viscosity might also be added. Once formed, the emulsion is stabilized either electrostatically with surfactants or sterically with natural or synthetic protective colloids. Wetting agents such as sodium polyphosphates and poly(acrylic) acids are added to disperse the pigments and extenders (Stoye and Freitag 1998: 126), and coalescing solvents are also added.

Emulsifiers, Surfactants, and Protective Colloids

The emulsifiers or surfactants that are used in the original catalytic emulsion polymerization process enable the hydrophobic polymer to be dispersed in water. Surfactants might be ionic, nonionic, or a mixture of the two. A huge variety of surfactants is available, and they are chosen to control the viscosity, particle size, and rate of reaction. A low-foaming and efficient emulsifying agent is needed to break the monomer into tiny globules and keep those globules in dispersion. The surfactant also acts as a protective colloid to prevent the globules from coagulating during polymerization, when their surface becomes slightly tacky; it also acts as a suspension agent to prevent the final polymer from settling in the can. The earliest surfactants were the soaps of salts of fatty acids, followed later by sulfated and sulfonated soaps. Anionic materials made good emulsifiers for this purpose, with nonionic ones being introduced in the late 1950s (Horn 1960: 105–6).

Some systems also require the addition of a protective colloid to help stabilize the emulsion. Protective colloids are water-soluble polymeric compounds of either natural or synthetic origin. They include natural gums, proteins, cellulose derivatives, polyvinyl alcohol, and acrylates (Naidus 1953). The protective colloid contributes to a range of properties. In addition to helping to prevent the particles from coagulating during polymer formation, it stabilizes the emulsion and helps keep the pigment particles in suspension. It also acts as a thickening agent, which ensures that the paint can be applied easily and flows properly during application. Although the protective colloid is one of the minor constituents of a paint in quantitative terms (typically no more than 1% to 2% of the emulsion solids and thus a fraction of 1% of the whole formulation), it plays a major role in controlling viscosity, stability, freeze-thaw stability, brushing, leveling, and good pigment dispersion. It also helps to prevent the emulsion from breaking under extreme conditions encountered during storage, transportation, and use (*American Paint Journal,* March 10, 1952, 57). The action of the protective colloid is indeed complex, evidenced by the fact that as late as 1953, although recognizing it as an essential component, manufacturers had little understanding of how exactly it worked or why it was necessary (Naidus 1953).

Protective colloids are not a requirement of all the emulsions discussed here: although essential for styrene-butadiene and styrene emulsions, they are not necessary in acrylics, and PVAc emulsions can be formulated with or without them. The type of protective colloid varies for each system: cellulose-based products such as hydroxyethyl cellulose, ethyl hydroxyethyl cellulose, and sodium carboxymethyl cellulose are most appropriate for PVAc paints (Warson 1972: 398) and could also be used for styrene-butadiene systems. However, casein and soybean protein, and to a lesser degree marine alginates, were more effective protective colloids for styrene-butadiene systems (Ronal and Weisberg 1954: 774). Although acrylics themselves do not require additional protective colloids, acrylate monomers were and are widely used as protective colloids and thickeners in both styrene-butadiene and PVAc systems ("The Markets" 1953: 47–49).

Although protective colloids perform a vital role in the stability of some emulsion systems, their permanent water solubility in the dried film is a considerable area of weakness. Because they are not absorbed into the polymer particles, they can be solubilized from the film on exposure to water and may confer haze and poor scrub resistance. It is therefore desirable to include the lowest possible quantity of protective colloid—no more than 1% to 5% by weight in solution (Erbil 2000: 150).

Biocides and Antimildew Agents

Many of the ingredients present in emulsion systems, and sometimes the resin itself, are nutrients for fungal and bacterial growth. Mold attacks the surface of the film—either in the can or once the film has dried—and flourishes in damp conditions. Bacterial attack occurs in the can, which causes the paint to putrefy, which in turn causes gassing. Minor cases of gassing might cause the can to expand, but a more serious consequence is explosion. It is therefore essential to add biocides and mildewstats to paints to prevent spoilage and mold growth: mildewstats prevent fungal growth, whereas biocides kill bacteria and fungus. As with all additives, it is essential to keep these to a minimum in order that they do not affect the properties of the paint adversely.

Although a great many biocides are available, the most effective for early emulsion paints comprised the aromatic mercurials, typically phenyl mercuriacetate, which was added to the order of around 0.1% by weight of paint. However, mercurials are not only toxic but also can cause discoloration if they react with sulfide-containing pigments, and high quantities were likely to be leached from the surface if exposed to damp conditions, which resulted in brown stains. Cheaper, less toxic, but also less effective biocides were the various substituted and chlorinated phenols (Warson 1972: 490). Zinc oxide pigment is also a mildewstat that has some fungicidal effect, so pigmentation with it was useful.

Antifoam Agents

Antifoam agents are essential ingredients that break up the air bubbles created during manufacture, transportation, and application. Foaming is a major problem for manufacturers and distributors alike and was arguably the biggest flaw of the early paints in particular, though it remains a problem today. Foaming is caused by the presence of surfactants, which themselves are essential for emulsifying and stabilizing the system. There are two types of foaming: the first produced during manufacture, and the second during application. The former comprises small, tight bubbles, whereas the latter comprises large, loose bubbles; both are virtually impossible to eliminate once formed (Scholl 1952c: 72).

If the paint has been allowed to foam, the air bubbles rise to the surface of the film and break, resulting in craters or pinholes on the surface. These craters are not only unsightly but also interrupt the coherency of the film and affect its ability to be washed: each crater represents an area of weakness, and rupture of the film can occur at these points when the film is exposed to water and cleaning agents. Furthermore, in early systems paints that had foamed during manufacture had a tendency to subsequently settle in the can, resulting in disgruntled customers complaining that the cans of paint they had bought were only half full. Foaming was a particular problem for styrene-butadiene systems, and additives such as pine oil and glycerol were recommended to reduce this tendency. The best solution to foaming was undoubtedly preventing its occurrence in the first place: manufacturers were warned never to allow the medium to freefall during manufacture and advised not to grind pigment into the resin, as the action of grinding caused excessive foaming (Scholl 1952c: 72–73).

Pigmentation

Pigmentation of emulsion paints is relatively straightforward, and a wide range of white and extender pigments are compatible with the resins. There is no benefit to using heavy metallic pigments in emulsion paint formulation: they are not only expensive but also tend to settle into a hard mass at the bottom of the can. Therefore, lighter pigments such as lithopone, zinc, and titanium white are included, along with colored organic pigments. By the early 1950s the most popular colored pigments were phthalocyanine blues and greens, Pigment Green B, toluidine yellow and red, BON red,[1] and maroon and hansa yellow ("The Markets: Dry Colors" 1952: 49). Acrylic resins in particular are easy to pigment and very stable, making them useful in the formulation of novelty metallic and fluorescent finishes that contain ionic pigments. However, styrene-butadiene required a little more care in pigmentation, as the presence of salts from either pigments or contaminants could break the emulsion.

Pigment manufacturers were quick to respond to the requirements of the paint industry: as grinding the pigment into the binder caused excessive foaming, manufacturers developed pulp colors, which were water-wettable pigment pastes or dispersions. In July 1952 DuPont was advertising a range of such pigments; the paste was designed to be mixed with hot water, then blended with the other ingredients, while the dispersion could simply be mixed in water (*American Paint Journal,* July 7, 1952: 37). In December 1952 DuPont was advertising an improved range, which had better resistance to efflorescence, nonreactivity in deep tones, and excellent washability, even in deep shades, the implication of course being that these were shortcomings present in the earliest pigments (*American Paint Journal,* September 15, 1952: 100–101).

The popularity of emulsion paints in the United States caused a sizable shift in the consumption of pigments: by the end of 1952 sales of organic color were booming, as were sales of lithopone and titanium dioxide ("The Markets" 1952: 35). Recognizing the popularity of emulsion paints, many pigment manufacturers entered this field in the early 1950s, and at the end of 1952 the largest volume of pigment consumption was the organic pigments used in emulsion paints ("The Markets: Dry Colors" 1952: 49).

Coalescing Solvents

Coalescing solvents are high-molecular-weight, slow-evaporating solvents that temporarily soften the polymer enough to allow it to form a coherent coating. The coalescing solvent then evaporates slowly from the film over a period of weeks or even months, leaving behind a harder, more durable film. It should be noted that the term *coalescing solvent* is unique to the paint industry, and the same products are referred to as *plasticizers* by the plastics industry. In addition, it is not uncommon to find *coalescing solvents* shortened to *co-solvents* by the paint industry, which misleads further: strictly speaking, co-solvents are secondary solvents added in small quantities to enhance the performance of the primary solvent. In the case of emulsion paints, co-solvents are added to increase the compatibility of ingredients such as viscosifiers with water. However, the role of coalescing and co-solvents is not always clear-cut. For example, the slow-evaporating alcohols such as glycols and carbitols that are frequently added to emulsion systems to prevent freezing might also act as coalescing solvents and even co-solvents if they are compatible with the polymer in the emulsion. However, any solvent that evaporates from the film is classed as a VOC, and as VOCs are gradually being eliminated, their presence in paints has been reduced in recent years.

Plasticizers

Plasticizers are essential to ensure that the polymer forms a coherent, flexible coating. Along with the temporary plasticizers (coalescing solvents) described above, which all emulsion systems require, additional plasticizers are necessary for the hard monomeric resins such as styrene, methylmethacrylate, and PVAc, all of which produce hard, powdery coatings without them. Hard resins can be plasticized either internally or externally: internal plasticization involves copolymerization with a softer monomer; external plasticization involves adding plasticizer once the polymer has formed.

A system is externally plasticized by first emulsifying the plasticizer in a solution of protective colloid. This emulsion is then blended with the aqueous resin emulsion. In the final resin, the plasticizer and the emulsion are united in the internal phase. However, the plasticizer migrates from the film over time and is either lost by evaporation or may be absorbed by the substrate. Although this initially results in a harder film, it is a serious weakness of externally plasticized systems, as it leaves the film weak and brittle over time. Although externally plasticized systems are in many ways inferior to internal ones, they do have the advantage of offering flexibility in formulation and significant cost savings for selected end-use applications (Green 1953).

Internal plasticization undoubtedly offers the most benefit. The plasticizer is copolymerized with the monomer and so cannot migrate from the film. In the case of styrene-butadiene the hard styrene monomer is softened by copolymerization with soft, rubbery butadiene; for the manufacture of acrylic emulsions, hard methacrylate monomer is copolymerized with soft acrylate monomer. The manufacture itself is simpler, as there is no secondary plasticizing process. Finally, improved water spotting, scrub resistance, and resistance to alkaline hydrolysis is usually observed, although it should be noted that these properties are also influenced by the chemical composition of the copolymer and surfactant system used (Warson 1972: 390). Plasticizing a system is always a matter of compromise: the polymer needs to be soft enough to coalesce during drying, but hard enough to produce a durable film.

Thickeners

Thickeners are added to emulsion paints to increase their viscosity, providing a gel-like consistency, which prevents the paint from dripping off the brush during application; this is especially useful for applications such as ceiling paints. Although protective colloids perform a thickening role, further compounds are added to increase this effect. Commonly used thickeners were purified clays such as magnesium-calcium-aluminium silicate complexes; these could be used with emulsions based on most standard emulsifiers and protective colloids and were suitable for use in vinyl ester copolymers and styrene-butadiene emulsions. A solution of sodium silicate might be used as a thickener for PVAc and acrylic resins: thickeners were added to the order of around 0.5% to 1.25% by weight of paint (Warson 1972: 474–75).

Extenders

Extender solids are an important addition to any emulsion paints, where they improve properties such as flow. In early emulsion paints they made up about one-third of the total pigment content, although today they are present in greater quantities. The hiding power of extenders is low, but they usually flat the finish a little and for this reason are sometimes excluded from glossier finishes. In the early 1950s the principal extenders were kaolin, magnesium and aluminium silicates, pyrphylite clay, diatomaceous silica, whiting, barytes, and mica; each offered different characteristics, depending on what was required by the manufacturer (Scholl 1952b: 66).

STYRENE-BUTADIENE PAINTS

Emulsion paints based on copolymers of styrene and butadiene were first introduced to the U.S. paint market in the late 1940s. They were used in the manufacture of interior wall paints and were also marketed for exterior use, although their durability in the latter application was poor. As with all emulsion paints, they were easy to apply, easy to clean up, and quick drying. They also had good alkali resistance and scrubbability and could be recoated easily: they thus proved immediately and immensely popular with the home user. Styrene-butadiene paints were not used in British or European markets due to the unavailability of the monomers.

Styrene-butadiene copolymers were initially used in the production of synthetic rubber, a vast industry in the United States during and after World War II. When the war ended in 1945, alternative outputs were sought for both the polymer plants and the polymerization processes. It was found that a resin suited for surface coatings could be produced if the ratio of styrene to butadiene was altered. Synthetic rubber comprised 75% to 80% of the soft, flexible butadiene and 20% to 25% hard styrene. The resulting copolymer was too soft to be used in surface coatings, but by reducing the amount of butadiene to 30% to 35% and increasing the styrene content to 65% to 70%, a serviceable coatings resin was produced (Scholl 1952a: 64). Coatings were also manufactured from plasticized polystyrene (see Polystyrene Emulsions, below).

Manufacture

Styrene-butadiene emulsions were made in the manner described in Overview of Emulsion Polymerization, above. Once polymerized, the emulsion was stabilized by a protective colloid such as casein, soybean protein, alginates, methylcellulose, and polyacrylates ("Forum on Latex Paints" 1953: 11). The most important of these in the early years were undoubtedly casein and soybean protein, which also provided body to the paint and aided dispersion; in the mid-1950s proteins were reportedly used almost universally for this purpose (Ronal and Weisberg 1954: 774). Indeed, the use of casein by the paint industry increased at a substantial rate once styrene-butadiene paints were introduced, with a gallon of paint containing about 3% of casein or another stabilizing agent ("The Markets: Miscellaneous Materials" 1952b: 39).

Sequestering agents such as polyphosphate were added to the order of about 0.05% to 0.1%. They helped to neutralize the action of metallic salts that were everywhere: in the raw materials, the containers, and the manufacturing equipment. Salts were a significant problem, as they caused the emulsion to break, a fault especially acute with styrene-butadiene emulsions ("The Markets: Miscellaneous Materials" 1952a: 41). By 1952 Dow was advertising an improved formulation of Latex 762 W, which had better freeze-thaw stability and reduced sensitivity to ionic contamination (*American Paint Journal,* September 15, 1952, 74–75).

Foaming was perhaps the major flaw in these paints, and various defoamers were used. These included silicones, higher alcohols, pine oil, and tributyl phosphate. Defoamers work by coalescing the air bubbles that form, which then burst on the surface of the liquid paint. However, it is important that pinholes, or "fish eyes," do not appear on the surface of the paint film as a result. Antioxidants might also be added to reduce embrittlement; these included phenyl-napthylamine types and phenol-formaldehyde inhibitors (Warson 1972: 429).

Attempts were made to improve the durability and scrubbability of styrene-butadiene paints by modifying them with alkyds, heat-bodied drying oils, and long-oil varnishes. The modifying component was pre-emulsified with a 5% to 10% nonionic wetting agent, which was then further emulsified with an alkali, blended with a colloid solution, and then added to the final paint. Modification was intended to improve adhesion, leveling, and stability, although it is evident that it could affect the durability of the film. One author contended that modification with an alkyd resin in fact offered no particular advantage, and tests concluded that its inclusion could actually impair the stability of the emulsion and decrease washability of the dried film (Scholl 1952c: 72). This was due to the fact that the alkyd was emulsified with an alkali soap to form the soaps of a fatty acid. The particle size of this soap was much larger than that of the synthetic polymer, and on evaporation of the water the soap existed as large globules in the film. Some of these were the straight soaps of fatty acids and were thus extremely water sensitive in the dried film. When the coating was washed, the globules would dissolve and cause film failure more rapidly than coatings formulated without alkyd. Alkyd-modified styrene-butadiene paints also took longer to harden than unmodified versions. However, they were certainly advertised by many companies, such as the Eaglo Paint and Varnish Corporation's Magic Satin, which was introduced in October 1952 ("New Products"

1952: 92). Today the addition of an alkyd resin is considered deleterious, although the practice continued until the 1990s.

The properties of styrene-butadiene paint could also be influenced by the monomer addition schedule. For example, greater washability could be achieved if a portion of the styrene monomer was not charged until late in the polymerization process. Therefore, paints with identical quantitative compositions could have quite different properties, depending on how they were manufactured (Fowler and Westerhoff 1952: 90). Styrene-butadiene latexes were usually supplied at between 45% and 48% solids and a pH of 9 to 10; therefore, pigments needed to be stable up to pH 10. The average particle size of the styrene-butadiene emulsion was 0.2 micron (Warson 1972: 428).

History

Styrene-butadiene latexes were originally developed as a replacement for natural rubber, when imports of rubber from Southeast Asia were cut off during World War II. Although resins suitable for use in coatings were not commercialized until the late 1940s, German, British, and U.S. patents describing copolymers containing small amounts of butadiene and large amounts of styrene had been filed periodically since 1930 (Ianni, Hess, and Mast 1951). But it was not until the mid-1940s that research into styrene-butadiene emulsions began in earnest.

One of the first commercially available resins was Goodyear's Pliolite S-3. It comprised a 15:85 butadiene-styrene copolymer, and its properties and applications were first described in 1946 (Borders, Juve, and Hess 1946). Dow's first binder specifically for emulsion paint, Latex 512-K, was introduced in 1948 and by 1949 was being widely marketed to paint manufacturers; the binder was marketed for use in interior flats and sealers and exterior paints. The tendency for the emulsion to break in the presence of polyvalent cations was listed as a weakness, indicating the severity of the problem that emulsion break posed (Ryden, Britt, and Visger 1949: 74).

Despite the immediate popularity of styrene-butadiene paints, there were factors that hindered their production in the early years, primarily that monomer production could not keep up with demand. In 1951 there was a deficit in the production of both styrene and butadiene. The reason for this was the fact that the government's synthetic rubber program absorbed the supplies normally used for resinous plastics and other applications, including emulsion paints. Compounding the shortage, the manufacture of styrene resins and plastics had been increasing at an annual rate of 50%. The shortage was alleviated in late 1951, when new capacity of both monomers came into production ("The Markets: Varnish Gums" 1952: 49). However, the lack of a steady supply of the emulsified polymer had prevented many companies from entering the field until this deficit was rectified. Furthermore, many manufacturers believed—erroneously—that special equipment and processes were needed to formulate paints based on styrene-butadiene. In an effort to address this misunderstanding, the *American Paint Journal* issued a series of articles in 1952 about the manufacture and use of styrene-butadiene paints, one of which focused on how to manufacture them in a plant intended for oil-based paints (Scholl 1952c).

By 1952 many U.S. companies had entered the field, but most manufacturers did not in fact make their own paints. Some bought the emulsified polymer from a manufacturer and pigmented it themselves; others bought paints ready-made, then labeled them as their own. Many companies produced latex bases that could be mixed with pigment and thinned with water by the manufacturer.

Unless very-low-cost coatings are required, styrene-butadiene emulsions are no longer used for the manufacture of decorative paints. They are still used, however, for applications such as paper coatings.

Properties

Styrene-butadiene paints gained immediate popularity with both professionals and retail customers. On a superficial level they were easy to apply and covered well, and equipment could be easily cleaned with water. On a chemical level they had excellent binding properties and adhesion to fibrous surfaces, they produced a tough and flexible film, and they had good resistance to alkalis, heat, abrasion, and water (Williams 1960: 27). Their aging characteristics and light stability were, however, poor: a clear film has a yellowish color, and the binder has a tendency to yellow further upon aging. Styrene-butadiene latexes are high polymers with high solids content. This gave the paint excellent nonpenetrating properties, meaning that it could be applied to many different surfaces without requiring special preparation. The resin thus found extensive use in undercoats or primer sealers and could also be applied over surfaces previously painted with oil paints.

Styrene-butadiene's nonpenetrating properties also enabled deep-tone wall paints to be produced for the first time. Previous attempts at dark tones with the available water- and solvent-based paints often resulted in uneven gloss, or "flashing," caused by uneven absorption of the paint by the substrate, a fault especially visible in dark tints. The nonpenetrating nature of styrene-butadiene cured this fault. The quick-drying nature of the binder and its low odor rendered it suitable for decorating public buildings such as hotels, schools, and hospitals, as the walls could be painted in the morning and the rooms occupied the same afternoon. This was a huge benefit over the previously used oil-based paints, which not only took time to dry but also had a strong odor.

The ease of application of styrene-butadiene paints was undoubtedly their main advantage, especially compared to oil-based paints, which dragged the brush. The excellent application properties ensured that they were immediately popular with the home user: indeed, these paints were the first to be geared specifically to the retail market. One author stated that the "paint it yourself" trend could, in some measure, be traced directly to the advent of styrene-butadiene paints (Scholl 1952c: 70). Furthermore, whereas traditional water paints were water sensitive, with poor washability, styrene-butadiene paints exhibited excellent water, chemical, and abrasion resistance and remained more flexible. Because the resin contained no ester groups, the paints were completely resistant to hydrolysis. This meant that they could be used on alkaline surfaces such as new plaster.

Although styrene-butadiene paints had many good qualities, they also had significant drawbacks. They had all the shortcomings of synthetic rubber and discolored and became brittle on aging. The paints also had poor freeze resistance, suffered emulsion breakage, had poor shelf life, and developed viscosity buildup in storage (Horn 1960: 139). But their most common fault, ironically, was caused inadvertently by their excellent application properties: because they could be brushed out with ease, inexperienced users tended to brush them out far too much, with the result that the paint was stretched beyond its covering capacity. Styrene-butadiene paints needed to be applied with a full brush and flowed onto the surface. The film in fact needed to be applied in a thicker, heavier coat than did oil-based wall paints, and with the exception of very pale tints, one coat sufficed to achieve coverage (Scholl 1952a: 70). Insufficient coverage for one-coat applications in white and pale tints was due to the fact that, like most polymeric binders, the binder itself becomes almost completely transparent once dry and so gives no opacity to the final film. Finally, although possessing lower odor than equivalent oil-based finishes, styrene-butadiene paints had an unpleasant smell (Blackley 1997: 378).

Although the chemical and water resistance of styrene-butadiene paints was generally high, their light durability was poor. The polymer is unsaturated, and as it oxidizes in the presence of UV light, it becomes harder and discolors in much the same manner as natural rubber. Although undoubtedly disfiguring, the discoloration and embrittlement developed slowly and was probably no worse than that exhibited by oil-based paints (Scholl 1952a: 65). The susceptibility of styrene-butadiene paints to UV radiation meant that they were more suited to interior applications, although they were marketed for exterior finishes as well.

Styrene-butadiene paints were certainly easier to apply than oil-based paints, but the high surface tension of the water present meant that leveling was not as good as the oil-based equivalents, although by 1952 this property had been improved dramatically. As with all waterborne systems, styrene-butadiene paints were susceptible to freezing if exposed to low temperatures. Manufacturers thus built into their paints resistance to freezing, and most were stable over several freeze-thaw cycles. Finally, the pigment range was restricted to those unaffected by alkalis, as the binder, with a pH between 9 and 10, was alkaline in nature.

Availability and Use

Styrene-butadiene paints proved immediately and immensely popular, largely because they filled a niche for a paint that was as easy to apply and use as traditional water-based paints but had durability, scrubbability, and recoatability similar to that of oil paints. Manufacturers were quick to recognize their potential, and by 1949 advertisements for them appeared with increasing regularity: Sherwin-Williams's Super Kem-Tone, introduced in the late 1940s, was based on styrene-butadiene copolymer, as was Glidden's Spread Satin, which was introduced around the same time. In January 1952 *Time* magazine reported that around fifty companies were manufacturing latex paints, and they expected the number to increase to over one thousand the following year (Mann 1952: 41). Indeed, in August 1952 the *American Paint Journal* stated that from the handful of companies manufacturing them on their inception four or five years earlier, there were now more than sixty, and they expected 50% of the approximately thirteen hundred manufacturers in the United States to be making them by the end of the year (Scholl 1952a: 68): these figures give some idea of the extent of the paints' popularity (see plate 18). By 1958 an estimated seventy-one million pounds of styrene-butadiene were used in the production of paint, with a projected capacity of around ninety-eight million pounds by 1964 (Williams 1960: xii). They are available to a limited extent today for very-low-cost interior decorative paints, although this market has largely been displaced by vinyl acetate–acrylic copolymer in the United States.

POLYSTYRENE EMULSIONS

The very low cost of styrene monomer in the United States encouraged manufacturers to experiment with it for many different applications, and in 1950 polystyrene emulsions were introduced to the U.S. paint industry (Warson 1972: 427). Styrene conferred many good properties to paints, but it also had some serious inherent faults that limited its use in the coatings field: namely, the tendency of styrene to yellow on exposure to UV radiation, and the need for considerable quantities of plasticizer. Plasticized styrene emulsions were nonetheless used for paper coatings, hot plaster seals, and, to a lesser extent, wall paints. They were especially useful where improved durability was required and found use in cellar and bathroom paints (Graves 1952: 78).

Manufacture

The conventional methods of emulsion polymerization described above apply to styrene polymerization also, and emulsions were manufactured by the batch or the delayed addition process. Styrene, initiator, and emulsifiers were charged in a kettle with a stabilizer such as the ammonium salt of vinyl acetate–maleic anhydride. As styrene is a hard, brittle monomer, plasticizer amounting to nearly half the weight of the polymer was added. The emulsions could be pre- or postplasticized, and plasticizers such as dibutyl phthalate, polyethylene glycol, and di-2-ethylhexanoate were used. The recommended method was to make a separate plasticizer emulsion based on an ammonium soap and blend it hot with the polystyrene emulsion. Styrene was notoriously difficult to plasticize, as the plasticizers were effective only in a very narrow range: too little, and the film acquired a "cheesy" appearance; too much, and the film became soft and rapidly lost strength (Warson 1972: 427). Although it was possible to postplasticize the styrene emulsion, for convenience and assurance of consistent quality, most paint manufacturers bought preplasticized emulsions.

Having bought the plasticized emulsion, the paint manufacturer would mix it with pigment ground in either casein or polyacrylate. In the case of polyacrylate-ground paints, the wetting agent, sequestering agent, sodium polyacrylate dispersion, and water were ground thoroughly, then the pigments were mixed in. One pass through a three-roller mill was recommended, followed by transference of the paste to a paint mixer, to which was added the polystyrene emulsion, casein dispersion, the remaining water, ethylene glycol, and trietheylene glycol. The process was the same for casein-ground paints, except that the casein was used to grind the pigments, and the polyacrylate was added at the final stage; however, due to putrefaction problems, the casein-ground paste had to be used immediately, while the polyacrylate version could be left for over a month.

The addition of glycols helped brushability and flow, and pine oil might also be added to prevent foaming and improve the paint's rheological properties. It was essential to avoid aeration of the polystyrene emulsion during manufacture, as air bubbles in styrene emulsions were slow to rise to the surface: as with all emulsion systems, it was far preferable to avoid the occurrence of foaming in the first place. The addition of protective colloids such as casein and soybean protein stabilized the emulsion, enabling it to withstand mechanical abuse and several freeze-thaw cycles (Graves 1952: 82–85).

Properties

Paints based on polystyrene emulsions had many good properties: they were quick drying, chemically inert, dimensionally stable, and completely resistant to hydrolysis. They were also nonpenetrating, even when applied to very absorbent surfaces, which, when combined with good alkali resistance, made them very useful in the manufacture of hot plaster seals (Graves 1953: 64). Styrene paints dry by evaporation of the water, followed by coalescence of the polymer: the film does not pass through a tacky stage and therefore can be recoated within minutes. Wall paints made from styrene dried within twenty minutes and became fully scrubbable within twenty-four hours. Although the film was moisture and scrub resistant, it allowed the passage of moisture through it, which reduced the risk of blistering if applied to a damp surface (Graves 1952: 78). Finally, unlike styrene-butadiene emulsions, which broke relatively easily, the styrene emulsion was very stable under mechanical stress and retained its viscosity for up to two years in the can. The emulsion could also be left in manufacturing equipment for long periods without deleterious effect, as long as it was not allowed to dry.

However, styrene does have some significant inherent drawbacks that limited its use in air-dry coatings and made styrene-butadiene or PVAc paints preferable in most applications: styrene is a hard, brittle monomer that requires nearly 50% of its weight in plasticizer in order to produce a film of sufficient flexibility: this amounts to far more than is required in, for example, PVAc resins. Styrene is also far more difficult to plasticize than PVAc, and the plasticizer is lost more easily from the film, by leaching, migration, or evaporation, which leaves the film weak and brittle. Furthermore, despite manufacturers' claims that the light stability of styrene was good (Scholl 1952a: 65), styrene yellows considerably on exposure to UV light. Finally, styrene had inferior leveling characteristics when compared to other systems. These faults meant that decorative paints based on styrene were nowhere near as popular as those based on copolymers of styrene and butadiene, although they were used extensively in industrial coatings (Graves 1953: 64).

POLYVINYL ACETATE

The development of coatings based on PVAc was carried out in Germany. The first solution coatings were introduced in 1930 and the first emulsified systems in the mid-1930s. However, they did not make an impact on the paint industry until emulsions became widely available in the late 1940s. PVAc has many good properties, and coatings based on it are mechanically stable, easy to apply, and easy to clean up; have excellent adhesion to wood, plaster, and cement; and are resistant to UV radiation and oxidation. Their water and alkali resistance is reasonable to good, which, along with their moisture permeability, makes them useful for sealing new and damp plaster. They have thus been used as binders for primer sealers, exterior masonry paints, and interior flat wall paints. PVAc is frequently used in combination with other monomers. Monomeric PVAc resins were and are more widely used in the United Kingdom and Europe than the United States, but from the mid-1950s it has been common to copolymerize PVAc with acrylic in the United States. PVAc remains the most widely used binder for interior decorative paints in Britain and Europe today.

Manufacture

The emulsion polymerization of PVAc involves polymerizing the vinyl acetate to a solids content of about 50% to 55% in the presence of 0.05% to 1.5% surfactant. The surfactant can be anionic, nonionic, or a mixture of the two, but it must be stable in acidic conditions, as polymerization occurs between pH 4 and 6. Protective colloids in the form of polyvinyl alcohol (PVAl) or cellulose derivatives are frequently included. It is possible to make PVAc emulsions with all surfactant, all protective colloid, or a mixture of the two. The amount of surfactant and protective colloid present is usually to the order of up to 3%, and a buffer such as sodium bicarbonate might also be added to control the pH (Feldman and Barbalata 1996: 92–93). The reaction takes between four and six hours and is run at temperatures between 65°C and 90°C. For paint use, emulsions cannot contain more than 0.5% free monomer, or odor and stability problems arise. Excessive free monomer also softens the dried resin film; therefore, free monomer must be removed (Davis 1954: 80).

Unmodified PVAc films are too hard and brittle to form a continuous coating, so in order to obtain a softer, more flexible film, internal or external plasticization is required. Only externally plasticized resins were available until the mid-1950s. The emulsion was modified with around 2% to 5% of a plasticizer such as dibutyl phthalate or tricresyl phosphate (Williams 1960: 32). However, externally plasticized films eventually become weak and brittle, as the plasticizer has a tendency to volatilize and evaporate from the surface of the film. In an effort to overcome this weakness, internally plasticized resins were introduced in the mid-1950s.[2] The resin was internally plastized by copolymerizing vinyl acetate with a softer monomer.

Copolymerized resins have much greater acid, alkali, and water resistance than externally plasticized versions, but they did not displace the latter, which continue to be widely used.

Coalescing solvents such as glycols and glycol ethers were also added, as were thickeners, wetting agents, pigment dispersants, mold inhibitors, and antifoam agents. By the mid-1950s PVAc films were also being modified with alkyds, with the intention of increasing the scrub resistance, freeze-thaw stability, and pigment-wetting properties (Davis 1954: 82).

To formulate a paint from PVAc emulsion, the pigments are ground into a paste with a wetting agent such as sodium hexametaphosphate, together with a nonionic protective colloid such as hydroxyethyl cellulose, methylcellulose, or ethyl hydroxyethyl cellulose. Polyelectrolytes such as maleic anhydride–diisobutylene copolymer, or polyacrylate and polymethyl acrylate salts, can also be used. The defoamers and antibacterial agents are added in the final mixing of the paint (Warson 1972: 381–82).

PVAc emulsions have only moderate pigment-binding power, so very high ratios of pigment to binder should be avoided, although this might be employed in cheap products such as ceiling paint, which were intended to simulate distemper. Although PVAc resins are compatible with most pigments, the pH of PVAc emulsion is slightly acidic, so basic pigments such as calcium carbonate should be avoided. If basic pigments are used, the resulting release of carbon dioxide as they react with the acidic resin can cause the can to expand or even explode. The main white pigment is rutile titanium dioxide, which comprises around 60% to 70% of the total pigment content. Extenders include wollastonite,[3] talc, china clay, barytes, and mica. Most of the pigments used in styrene-butadiene systems were satisfactory for PVAc paints, but deep-tone colors and tinting colors needed to be carefully selected to avoid flooding, streaking, and shade variations. As the paints are slightly acidic, a suitable container must be chosen to ensure adequate shelf life (Davis 1954: 82).

History

Vinyl acetate was first synthesized in 1912 and first polymerized in 1914. By the 1920s the production of vinyl acetate and vinyl alcohol monomers was being conducted in Germany, the United Kingdom, and the United States (Lewin and Pearce 1998: 280). Much of the development of PVAc coatings and adhesives was carried out by the German company Hoechst AG, which in 1930 introduced Mowilith H and N resins. These were solvent-borne PVAc resins, marketed as colorless, lightfast, and tough. They were soluble in cheap solvents and were thus used in lacquers as a replacement for nitrocellulose where good lightfastness was required.

Hoechst AG was also responsible for the development of emulsified PVAc resins and in the mid-1930s introduced Mowilith D and D32.

These were PVAl-based polyvinyl acetate resins;[4] the polyvinyl alcohol acted as a protective colloid and also helped to plasticize the resin (Erbil 2000: 6). The manufacturing technique, which consists of polymerizing the pre-emulsified monomer with peroxide initiators and oleate salts, is still used in many plants today (Olagoke 1997: 255). Mowilith D32 was the plasticized version of Mowilith D and was plasticized with 20% dibutyl phthalate and 13⅓% tritolyl phosphate (Warson and Finch 2001: 722). For use in decorative paints, Mowilith D and D32 were employed in equal proportions, resulting in polymers plasticized to about 10%. The pigment pastes were prepared separately and mixed just before use. Indeed, as late as 1953 manufacturers were providing PVAc paints either ready to use or in paste form (Snelling 1953b: 55). The resins were compatible with most of the standard pigments, and the paints were marketed for interior and exterior use. However, the emulsions had poor gloss, so only flat finishes could be produced, and the presence of PVAl also made them susceptible to hydrolysis. These PVAc emulsions were investigated by the Allied teams of scientists who visited German factories at the end of World War II, and the first paints based on these recipes were introduced in Britain and the United States in 1948 (Erbil 2000: 6). The manufacturing process for Mowilith D was the fundamental type operated in many plants until at least the early 1970s.

The main disadvantages of PVAl-emulsified resins were poor gloss, high viscosity, and a tendency to re-emulsify, although they did have reasonable freeze-thaw stability. Despite these shortcomings, the use of such resins in Germany increased dramatically during World War II, largely due to the acute shortage of vegetable oils traditionally used in paint manufacture. Resins of this type were also being manufactured in the United States and Canada during the 1930s, although Britain did not enter the market until the 1940s (Dulac 2007: 171).

In the mid- to late 1940s Shawinigan Laboratories in Canada developed improved emulsions. Shawinigan used gum arabic and hydroxyethyl cellulose in place of PVAl as the protective colloid, included anionic or nonionic wetting agents, and prepared the emulsions using the delayed addition process (Olagoke 1997: 255). The resulting resins were more stable and had vastly improved properties, including lower viscosity, better gloss, better water resistance, better resistance to blanching, a considerably smaller particle size (0.3–1.0 µm rather than the 1–3 µm of PVAl types), and less tendency to reemulsify. These resins were plasticized externally, and the plasticizers could simply be stirred directly into the PVAc emulsion, with optional use of heat.

However, any externally plasticized system has the inherent problem of loss of plasticizer. In order to overcome these problems, vinyl acetate monomers were internally plasticized via copolymerization with other monomers. Introducing a copolymer not only plasticized the resin but also increased the resin's resistance to acid and alkaline hydrolysis and improved its weather resistance. Copolymers of vinyl acetate as a basis for emulsion paints were introduced around 1953–54.

The type of comonomer chosen depended on availability, cost, chemical properties, ease of polymerization, and plasticizing ability. The earliest copolymers comprised vinyl esters of maleic and fumaric acid (dibutyl maleate and dibutyl fumarate) and acrylate esters. From the late 1960s and early 1970s, derivatives of itaconic acid and ethylene were also used. The copolymer made up about 15% of the total resin (Warson 1972: 196, 368). From the mid-1950s vinyl acetate was also copolymerized with acrylic esters to produce a binder that is widely used in the United States for wall paints today. However, it is evident that there were problems with these early vinyl acetate–acrylic resins, as it would appear that some manufacturers did not understand fully the polymerization process. All PVAc copolymer emulsions must be manufactured using the delayed addition process. This means adding the monomers at a rate equivalent to the speed of the reaction, which ensures that no buildup of the slower-polymerizing vinyl acetate occurs. Many manufacturers did not take into account the different copolymerization rates of acrylic esters and vinyl acetate. As a result, many advertised "acrylic copolymers" were in fact either practically all acrylic or all PVAc. Although an apparently clear film is produced when dried, the physical properties of the resulting resins were quite different from true copolymers, and their flexibility was often seriously impaired (Warson 1972: 198–99).

Despite the improved properties of copolymerized resins, their use did not become widespread until the mid- to late 1950s, and both internally and externally plasticized resins are still widely used today.

Properties of Paints Based on Polyvinyl Acetate

PVAc resins are clear, colorless, odorless, nontoxic, chemically inert to UV light and oxidation, grease-proof, and mildew-proof. The polymer is soluble in organic ketones, esters, chlorinated hydrocarbons, aromatic hydrocarbons, and alcohols and is insoluble in water, aliphatic hydrocarbons, fats, and waxes. Emulsion paints based on it have many excellent properties that have ensured their continued popularity: paints based on PVAc resins are mechanically stable, easy to apply, and easy to clean up; have excellent adhesion to wood, plaster, and cement; are resistant to UV radiation and oxidation; and exhibit reasonable water and alkali resistance.

PVAc resins are inherently suitable as primer sealers and undercoats for oil-based paints: they are nonpenetrating, have the ability to seal or bridge small gaps, and have good adhesion. They also have better oil resistance and resistance to the solvents found in oil-based paints than either styrene-butadiene or polystyrene paints. Finally, they are very quick drying, which enabled the user to apply two coats in one day, representing a considerable savings of time ("Forum on Latex Paints" 1953: 10). PVAc resins are reasonably scrubbable, and although it takes several days for them to achieve maximum hardness, this marked a considerable improvement on paints that dried by oxidation and took many days or even weeks to achieve maximum hardness.

The main weaknesses of PVAc resins are their hard, brittle nature (which makes plasticizer necessary), their sensitivity to water, and their susceptibility to permanent damage from freezing in the can (Williams 1960: 33). The alkali resistance of PVAc resins varies and is to a great degree determined by the composition of the emulsion system: internally plasticized resins exhibit greater water, acid, and alkali resistance than externally plasticized ones. PVAc resins containing PVAl cannot be applied to highly alkaline surfaces due to hydrolysis of the resin and subsequent leaching away of the PVAl: most copolymers reportedly suffered from similar problems (Warson 1972: 383). Although all water-thinnable paints contain water-extractable material, early PVAc resins contained appreciable amounts and were considerably more water sensitive than styrene-butadiene and acrylic paints. Most of this material came from the basic resin emulsion binder, and some binders of this type contained about 5% of readily extractable material. This was a significant problem for manufacturers, as these coatings tended to water-spot excessively and had poor scrub resistance and wet adhesion to sealed surfaces. This made them unsuited for applications such as kitchen and bathroom paint. However, by the mid-1950s techniques had been developed for making emulsifier-free PVAc resins with virtually no extractable material (Scheffele and Tompkins 1959: 114).

The durability of PVAc resins could be further improved via the addition of alkyd resins or by copolymerization with other monomers. Common copolymers included acrylates and vinyl esters of branched chain fatty acids; one example of the latter were Veova resins, which were developed in Germany during World War II. Superior chemical and water resistance was achieved by copolymerizing vinyl acetate with about 20% of vinyl benzoate (Warson 1972: 401). It is possible that the German PVAc paints written about in 1954, which had undergone seven years of exposure testing without loss of adhesion, color, or durability, were of this type (Davis 1954: 82).

Availability and Use

There was a significant difference in the availability and use of PVAc emulsion paints between countries. They were used extensively in Germany throughout World War II, when vegetable oils were in short supply. Canada's early and large acetylene industry ensured that the vinyl acetate monomer was widely and cheaply available, and PVAc paints were introduced to the Canadian market by Shawinigan Chemicals in the mid-1940s. Although the earliest decorative finishes based on PVAc reportedly appeared in the United Kingdom in 1946–47, they did not make an impact on the British paint industry until the improved emulsions appeared in 1948–49 (Warson 1972: 367–68) and in fact do not appear in British trade literature until late 1951. It is not until 1952–53 that large numbers of advertisements and discussions of their properties appear. Although PVAc emulsions were widely used in the United Kingdom and Europe, they did not gain popularity in the United States until the mid-1950s, when vinyl acetate monomer became financially competitive ("Vehicle Manufacturers" 1954: 9–10).

In the United Kingdom PVAc paints quickly became known as plastic emulsion paints, which were available ready to use or in paste form that could be thinned with water (Snelling 1953b). The name given to them is confusing, as "plastic paints" had previously been used to describe textured coatings that bore no relation to paints based on synthetic polymers. PVAc paints were featured heavily by many manufacturers at the 1951 Building Exhibition at Olympia, London ("Plastic Emulsion Paints" 1952: 52). In 1952 a vast number of products were introduced to the British market, all of which were described as moisture permeable, quick drying to an eggshell finish, and with low odor and good adhesion to a range of substrates.

Although PVAc resins were introduced at a slightly later date in the United States, by the early 1950s large numbers of advertisements began to appear in the press for PVAc-based primer sealers (*American Paint Journal,* November 10, 1952: 103). Due to the widespread availability of styrene-butadiene systems, their use in the United States was initially confined to primer sealers, undercoats for oil-based wall paints, and exterior masonry paints—applications for which they were superior to styrene-butadiene systems. However, as the cost of PVAc decreased throughout the 1950s and 1960s, it was used increasingly in U.S. decorative paints, often copolymerized with acrylic: in 1958, seventeen thousand tons (dry weight) of PVAc was used in paints, and by 1965 this figure had risen to about fifty thousand tons (Warson 1972: 368). Many of today's emulsion paints in the United States comprise PVAc copolymerized with acrylic.

By 1960 nearly all emulsion paints in the United Kingdom were based on PVAc, and both internally and externally plasticized versions were available. The resulting paints were moderately durable and relatively cheap. In fact, as one author pointed out, greater durability was not always necessary, as decorative films tended to get repainted because they were dirty or an unfashionable color rather than due to film failure ("Higher Methacrylates" 1960). If film failure did occur, it was usually caused by inappropriate use of the paints. Just as had occurred with nitrocellulose brushing lacquers, overly optimistic claims were made for PVAc paints when they were first introduced ("Institute Brains Trust" 1952: 67). Users interpreted "good adhesion" as a license to paint over any old surface, whereas in fact the nonpenetrating nature of all emulsion paints prevents them from adhering to any chalky surface, and attempts to coat old distemper with PVAc resulted in the inevitable flaking and peeling. Misuse notwithstanding, the advantages of PVAc coatings by far outweighed their disadvantages, and their popularity continued to increase, until by 1960 they represented virtually the whole U.K. market for emulsion paints, which amounted to about ten thousand tons per year of dry polymer ("Higher Methacrylates" 1960).

ACRYLIC PAINTS

Acrylic resins are made up of esters of acrylic and methacrylic acid. Depending on the type and ratio of monomers used, acrylic resins can be cast into hard sheets, used as molding powders, or formulated into a flexible resin for use in adhesives and coatings. Solution acrylics were first introduced in 1927; emulsified versions, in the 1940s. The first household paints based on acrylic emulsions were not introduced until 1953, but largely due to the expense of the monomers, they did not enjoy widespread success until the 1960s. Paints based on acrylic resins have excellent properties, including good color and color retention, excellent water and alkali resistance, stability, and good scrubbability. Emulsified resins for household paints comprise a copolymer of about two-thirds acrylate and one-third methacrylate.

Acrylics were and still are more expensive than either styrene-butadiene or PVAc, and they thus tend to be used in speciality paints where greater durability is required, such as exteriors and in kitchens and bathrooms. In the United States acrylics are copolymerized with vinyl acetate to produce flat wall paints. Acrylic emulsions are also used extensively in the manufacture of artists' paints.

Manufacture

Acrylic esters are vinyl-type monomers, in that they contain the functional group $CH_2 = CRO_2R$. When an acid or carboxyl group is joined to the vinyl radical, along with an H or CH_3, methacrylic or acrylic acids are produced: these are the starting materials for all acrylic resins. Esterification of the acrylic acids with alcohol substituents yields a large group of acrylic esters (Horn 1960: 1–2). Nearly all of the acrylic monomers can be readily polymerized in emulsion form and copolymerized with each other. Acrylic emulsions are thus straightforward to manufacture, and the resulting emulsions have good stability.

Early acrylic emulsions for use in paints comprised a copolymer of 62% to 64% ethyl acrylate, which is polymerized with 1% to 2% methacrylic acid, with the remainder being methyl methacrylate. Polyelectrolytes were added to help stabilize the emulsion and also improved freeze-thaw stability and pigment wetting and increased the gloss of the film (Warson 1972: 416). Despite the presence of acids, the alkali resistance of acrylics is much higher than that of vinyl ester polymers, and the methacrylate units in the monomer are very resistant to hydrolysis. Acrylic emulsions have a very fine particle size, less than 0.2 micron, so in order to maintain emulsion stability in these early emulsions, comparatively high quantities of surfactant were required, about 7% to 8% of the monomers; far less surfactant is required today. Blends of ionic and nonionic emulsifiers were also added, depending on the end use. Acrylic resins do not require the addition of protective colloids. Acrylics have a solids content of 40% to 50%, and the pH is adjusted to around 8.0 to 8.5; historically, ammonia was used for this purpose, although it is considered too odorous today.

Acrylics also require conventional additives such as antifoam agents and preservatives. Coalescing solvents are less necessary than with PVAc resins but may be included if the coating is to be applied to a porous surface or for coatings with high pigment volume concentration ratios (Warson 1972: 416–18). Styrene and sometimes acrylonitrile could be added to harden the film, increase its chemical resistance, and also make the paints cheaper. However, even in small quantities, styrene tended to cause yellowing if the film was exposed to UV radiation, while acrylonitrile not only causes problems with wetting and adhesion but is itself a hazardous chemical that requires careful handling.

History

Acrylic acid was first synthesized in 1843, and by 1900 most of the common acrylates were known (Horn 1960: 10). The research and development into acrylic resin emulsions was conducted exclusively by the Rohm & Haas Company, and the discussion of acrylics is thus focused on this manufacturer. Rohm & Haas was established by Otto Röhm and Otto Haas in Esslingen, Germany, in 1907; in 1909 Haas moved to Philadelphia and established the U.S. branch of the company. Röhm had been interested in acrylic resins since his student days. Indeed, his 1901 Ph.D. dissertation had investigated acrylic acid and its esters and polymers, and in the same year he had prepared solid transparent polymers of acrylic acid derivatives (Horn 1960: 10). As early as 1915 Röhm had secured a German patent that suggested polyacrylates as a substitute for drying oils in paints and varnishes.[5] And by 1926 Röhm and his staff were close to synthesizing acrylic acid in commercial quantities. In 1927 they discovered the first commercial application for two of the simplest acrylic esters: methyl and ethyl acrylate. The soft, rubbery polymer was used as an interleaving layer for Luglas safety glass, which was introduced to the German auto industry in 1928. By 1931 Haas had set up his own acrylic laboratory in the United States, and in that year the company introduced its own version, Plexigum, for the U.S. safety glass market. The initial reception of Luglas and Plexigum was reportedly good, but the Depression caused a steep decline in use as sales of automobiles fell dramatically (Hochheiser 1986: 36–37). Although profitable on their introduction, the acrylate-based interleaving layers used in Luglas and Plexigum had inherent problems: they were soft and rubbery and could not be cut to shape easily, meaning that they had to be manufactured to the exact size required; they also performed poorly at low temperatures. By the late 1930s they had been superseded by polyvinyl acetate and polyvinyl butyral films for this application, which not only had superior properties but were less expensive as well (Hochheiser 1986: 56).

At the same time, methyl and ethyl acrylates were also formulated into coatings. In 1927 Rohm & Haas introduced its Acryloid resins, based on polymethyl acrylate. These early resins were available in solution form only and were used for industrial applications such as coatings for metals, automobiles, and signs (Horn 1960: 10). Acryloids were thermoplastic resins and dried by evaporation of the solvent: the films thus resoftened at high temperatures and remained permanently soluble in organic solvents. The coatings required baking to achieve full adhesion, hardness, and gloss (Rohm & Haas 1953: 1).

Although solution acrylics had many good properties, including excellent water resistance, they suffered defects that hindered their handling and use. Solution polymers must have low molecular weight to ensure solubility in solvents, but low molecular weight equals low solids content, which makes the films thin. Although it is possible to increase the solids content, this makes the solution very viscous and difficult to handle. The inclusion of solvents is also problematic: not only is their presence both a fire and a health hazard, but it represents a considerable expense, as the solvents are either lost through evaporation or require costly solvent-recovery systems (Horn 1960: 101). Emulsifying the resin in water overcame these problems.

The next notable step in the development of acrylic resins was the discovery of polymethyl methacrylate (PMMA), which was first synthesized in 1931. PMMA is a hard, brittle polymer, which can be copolymerized with the softer acrylates to produce resins with a complete range of hardness: from soft and rubbery to hard and brittle. Undoubtedly the most important commercial application for PMMA was in cast sheet form, where it was and still is used extensively for the production of lightweight plastic items. It is in this form that acrylic resins are perhaps most commonly known: as Perspex in the United Kingdom (initially manufactured by ICI) and Lucite (DuPont) and Plexiglas (Rohm & Haas) in the United States.

The technology for sheet casting PMMA was developed in the United Kingdom in the 1930s, primarily by ICI (Barron 1945: 491). ICI introduced Perspex in the United Kingdom in 1934;[6] Rohm & Haas introduced Plexiglas to both Germany and the United States two years later (Hochheiser 1986: 59). The consumption of acrylic sheeting increased dramatically during World War II, where it was used for fighter plane cockpit canopies and gun turrets. Acrylic was especially suited to this application, as it is not only lightweight with exceptional clarity but also can be heat-formed to produce curved shapes. During the war years the growth in sales of acrylic sheeting was staggering, with nearly all of it for military use. In the United States sales of Plexiglas grew from $5.5 million in 1938 to $43 million by 1944 (Hochheiser 1986: 65). In the United Kingdom production rose from 455 tons in 1940 to 6,000 tons in 1944.[7] One 1945 author prophesied that once the wartime demand fell off, the raw materials would become widely available and at lower cost, thus opening avenues for new uses (Barron 1945: 491).

At the cessation of war the demand for acrylic sheeting fell considerably. However, by that time the usefulness of the plastic had been recognized, and it rapidly became popular for applications as diverse as eyeglass frames, umbrella handles, and dentures, to name but a few. Moreover, by the late 1940s it became available in emulsified form, making it more attractive to the coatings industry.

Acrylic emulsions had certainly been investigated in Germany during the war, and although evidently far superior to existing PVAc resins, they found limited use, largely due to the high cost and lack of availability of the monomers (Warson 1972: 369). However, it was not until the late 1940s that Rohm & Haas introduced the first acrylic emulsions, Rhoplex and Primal, for finishing textiles and leather, respectively. In 1949 Rohm & Haas was advertising Rhoplex emulsions for use in the textile, paper, paint, and ink industries. The three Rhoplex resins were supplied as 40% solids and comprised WN-80 for coating fibrous surfaces, FRN as a saturant for porous materials or as a plasticizer for less flexible Rhoplex resins, and WN-75 for emulsion paints (*American Paint*

Journal, October 31, 1949, 68); however, they appear not to have been widely adopted for the last application at that time.

Rohm & Haas's scientists had in fact recognized the potential of acrylic emulsions for house paints around 1947, but the proposal had been rejected and the chemists instructed to focus their attention on textile emulsions instead. But by 1949 Haas realized that the retail house paint market was huge and began investigations into the use of acrylic emulsions for this application. This research was inaugurated by other factors also: first, in 1948 Rohm & Haas had merged with Resinous Products, a company that specialized in coatings and thus had the necessary expertise to formulate acrylic house paints; and second, the company was looking for an outlet for acrylate monomers. At that time the vast majority of acrylic resin was being used in sheet form, molding powders, or oil additives, all of which were based on methacrylate. Although acrylates were sometimes added to these products as a softening monomer, their use was otherwise restricted to the highly specialized Rhoplex textile finishes and Primal leather finishes, which had a relatively small market share (Hochheiser 1986: 97–98).

Although Rohm & Haas had devised an improved method for production of acrylate monomer, it was not economical to build a new plant unless an increased market for acrylates could be found. In 1949 Haas, seeing that house paints could be a huge consumer of acrylates, instructed that a new acrylate monomer plant be built; the plant was built in Houston, Texas, and became operational in 1953 ("Rohm & Haas Announces" 1953: 44). The first binder formulated specifically for use in house paints, Rhoplex AC-33, was introduced in 1953.[8]

In 1953 the Rhoplex family of emulsions was marketed to the paper, rubber, electrical, and textile fields as coatings, adhesives, and impregnants. All were characterized by extremely small particle size, and some were stable at widely varying pH levels, with excellent resistance to highly ionized salts; the latter property was especially important in the formulation of emulsion paints, as emulsion break caused by the presence of salts was a problem for styrene-butadiene paints in particular. Rhoplex resins could be used alone or with a variety of existing latexes. For example, the soft, flexible acrylate emulsion could be used as a permanent, nonmigrating plasticizer for PVAc films (Rohm & Haas 1953: 12).

It was immediately apparent that acrylic polymers were far superior to other paints based on synthetic polymer emulsions. They had good color and color retention, better scrub resistance, good flexibility, excellent storage stability, and resistance to freezing. They were quick drying, so they could be recoated within about one hour, had low odor, and could be applied to damp surfaces. They also had excellent resistance to hydrolysis, so they could be used on new plaster and cement. Acrylics overcame the problems inherent in both PVAc and styrene-butadiene paints in that they were simple to formulate, easy to manufacture, and very stable. The paints did not require additional protective colloids and were chemically stable in the presence of high salt concentrations and basic pigments such as calcium carbonate ("The Markets" 1953: 47–49). However, despite their good properties and the fact that they were far easier to formulate than existing systems, they had drawbacks that, along with their high cost, meant that they took some time to become established in the field. The defects of these early resins included poor brushing, low gloss, lack of adhesion, and visible brush marks; the exterior durability was also questioned (Hochheiser 1986: 99).

Although well established by the late 1950s, Rhoplex AC-33 was not one of Rohm & Haas's major products, and it was not until the technical qualities of the emulsions were improved that manufacturers became convinced that their benefits outweighed their defects. Initially, improvements were attempted by modifying the resins with additives—an unsatisfactory solution, as increasing the additives also increased the complexity of formulation. Scientists therefore looked at more fundamental research to solve the problems.

The growth of Rhoplex was the result of many years' research and the introduction of a series of improved emulsions. But even a decade after their introduction acrylic emulsions were still not commercially successful, to the extent that Haas and other senior managers in the company seriously considered abandoning the research altogether. Over the next twelve years, however, Rohm & Haas scientists worked on the resins, bringing out a series of better versions: AC-34, AC-22, AC-388, AC-490, AC-507, and AC-64. This research paid dividends. The flow and leveling of interior paints were improved, as were the gloss, hardness, and adhesion of exterior paints. As a result, large gains in sales were achieved, and by 1973 Rohm & Haas had the largest market share of interior semigloss paints. By 1978 acrylic emulsions had practically replaced the oil-based paints for the domestic market. In 1975 the company sold more than $58 million worth of acrylic emulsion paints in the United States alone, and it became the company's best-selling domestic product line (Hochheiser 1986: 138). By that time Rohm & Haas was not alone in monomer and emulsion production: beginning in the late 1950s many others had entered the field. By 1960 numerous companies were producing the monomer and manufacturing emulsions; the extent of use and availability of these resins is discussed below.

Properties

Paints based on acrylic resins are far superior to those based on PVAc and styrene-butadiene in terms of their durability and ease of manufacture. They have good color and color retention and excellent resistance to acids, alkalis, alcohols, water, oils, hydrocarbons, and chemical fumes. Because they have low reactivity, they can be used with basic pigments and for novelty finishes, including fluorescent coatings. They are straightforward to manufacture and easy to pigment, and the emulsions are stable. They are also very versatile and can be copolymerized with many other monomers. Acrylics have much greater resistance to hydrolysis than do PVAc resins and have greater elongation at break than do PVAc resins of equivalent hardness; this makes them especially suited to exterior paints, which are subject to extreme weathering.

The ease with which acrylics can be manufactured is also a great bonus: all that is required is mixing the pigment dispersion with the resin.

Furthermore, the emulsion itself is exceptionally stable, does not putrefy, and can withstand several freeze-thaw cycles. The paint is easy to apply with a brush or roller and dries so quickly that it can be recoated within minutes. The scrub resistance is not only good but also develops after just a few hours rather than the few days that it takes styrene-butadiene and PVAc paints. Furthermore, the coating does not become brittle or discolored with age, and dirt can be removed relatively easily from its surface (*American Paint Journal,* February 23, 1953: 17–20).[9] The extremely rapid drying time made acrylics popular with contractors: although more expensive than alternative systems, it was possible to apply the priming layer and the second coat in rapid succession, negating the need to move the scaffolding between coats, thus saving time and labor costs (*American Paint Journal,* February 1, 1954, 16–17).

However, despite the undeniably good properties of house paints based on acrylics, they made little impact when first introduced, which was due in part to their expense: manufacturers were understandably reluctant to invest in expensive systems that had not yet been perfected or adequately tested when serviceable alternatives were available, at a considerably lower cost. Furthermore, the paints suffered a number of problems and were inferior to the alkyd-in-water emulsions that were their primary competitors for durable paints in the early 1950s. The new acrylic coatings had poor flow and leveling, resulting in rough coats with visible brush marks, and the gloss and adhesion were also poor. Indeed, the lack of adhesion meant that they could not be used for coating wood. Consumers were skeptical about their exterior durability, something that Rohm & Haas attempted to rectify by painting the exterior of one of its own scientist's houses in 1956. This reportedly led to several million dollars worth of sales, as consumers recognized the benefits of acrylics for this application (Hochheiser 1986: 99).

The small particle size of the emulsion was also the source of problems. Because of the large surface area created by small particles, nearly all the surfactant present was being used to stabilize the emulsion; this made wetting the pigment difficult, as the interaction between the colloid in the pigment paste and the surfactant in the emulsion caused coagulation. Furthermore, the small particle size caused the emulsion to drag when brushed (Warson 1972: 417). Throughout the 1960s and 1970s the company worked hard to improve the emulsions, and by the late 1970s acrylic finishes had all but replaced oil-based wall paints for the domestic market. They remain popular for interior and exterior household paints, especially where durability, moisture, and alkaline resistance are required.

Extent of Use and Availability

Acrylic resins are extremely versatile and are used for a wide variety of applications. They have been cast into sheets and molded from powder into various shapes, the monomer has been added to oil and hydraulic fluid, and polymers have been used in solution and emulsion form as adhesives, saturants, and coatings; the monomers have also been used as thickening agents, plasticizers, and protective colloids for other aqueous paint systems.

Although undoubtedly far superior to equivalent water-based paint systems, acrylic emulsions for house paints gained little headway when they were first introduced in 1953. The use of acrylic paints was far more prevalent in the United States and Europe than in Britain, and they remain more popular in the former regions today. This was due to both the lack of monomer availability in Britain and the reported conservatism of the British surface coatings industry, especially on the consumer side (Martin 1947: 64).

Although Rohm & Haas conducted all the initial research into acrylic coatings and was the first to introduce them, many other companies entered the market once their potential had been recognized. In 1960 ICI published an internal report on an investigation of the extent of use of synthetic polymer emulsions in the U.S. house paint industry that offers useful insights (Williams 1960). By 1960 many companies had established their own monomer plants. Captive use of the monomers was high in that most of these companies produced their own molding powders, cast sheets, lacquers, oil additives, and emulsions from their monomers. Along with Rohm & Haas, Union Carbide and Celanese were producing acrylates by 1960, with Dow Badische due to enter the market shortly. Methacrylate was manufactured by Rohm & Haas and DuPont only at that time, although two other companies were in the process of starting methacrylate production. Acrylic emulsions were offered to the U.S. market by at least eleven suppliers by 1960,[10] with others considering the move. However, Rohm & Haas was not only by far the largest manufacturer of acrylic emulsions at that time but also the leading supplier of knowledge of acrylic emulsions to the paint industry.

In an effort to make acrylics cheaper, many of these companies modified the resins with cheaper monomers such as styrene or PVAc. For example, Monsanto introduced a styrene-acrylic binder (Lytron 680), which was gaining increasing acceptance due to its "adequate" performance and lower cost. By 1960 Celanese had an acrylate coating containing vinyl acetate emulsion and was test marketing some acrylic emulsions. Dow had a new acrylic latex aimed at the paint market. Union Carbide was strongly promoting 2-ethylhexyl acrylate as a comonomer in vinyl acetate paint emulsions. DuPont had also recently introduced acrylic-based exterior paint, which was thought to contain methyl methacrylate and ethyl acrylate (Williams 1960: 35–49).

The popularity of acrylic emulsions for paints is reflected in the sales figures: initial sales of Rohm & Haas's Rhoplex AC-33 were small: $1.4 million in 1954, rising to $6.1 million in 1958. And of the 320 million gallons of house paints manufactured in the United States in 1958, only 2% were acrylic, with 18% comprising other water-based paints, and the remainder taken by oil-based paints. However, over the years the market for Rhoplex AC-33 remained small but increased steadily each year: in 1960 sales totaled $7.6 million; in 1965, $8.5 million; and by 1970, $24.6 million (Hochheiser 1986: 137).

COMPARISON OF THE PROPERTIES OF PVAC, STYRENE-BUTADIENE, AND ACRYLIC PAINTS

Paints based on polymeric emulsions are versatile coatings that possess many good properties. They have low viscosity at relatively high solids content and thus offer good coverage in a single coat; they are nonpenetrating and capable of bridging small gaps, which makes them suitable as primer sealers; they are moisture permeable and alkali resistant and thus can be applied to damp plaster; they are scrubbable and easily recoated; and finally, they are nontoxic, easy to apply by brush, roller, or spray, and easy to clean up. They do, however, have some disadvantages: their nonpenetrating properties prevents them from adhering to chalky surfaces, so sound surface preparation is a necessity; they have low levels of gloss; and the emulsion systems themselves can be delicate and susceptible to foaming (Williams 1960: 154).

The storage stability of all the binders is good, although the pH of vinyl acetate polymer and copolymer tends to fall over time, resulting in a reduction in viscosity. Styrene-butadiene, PVAc, and acrylate emulsions all exhibit good film coalescence, but styrene's need for large quantities of plasticizer to ensure adequate film integration on drying meant that its use declined quickly.

All the binders produce films with good clarity and continuity. Properly formulated resins based on acrylic esters and/or vinyl acetate produce water-white films with excellent clarity, but styrene-butadiene has a yellowish hue, which becomes more pronounced as the film ages. The discoloration is caused by the styrene monomer and by oxidation of the butadiene; this oxidation also causes embrittlement of styrene-butadiene films, and they deteriorate in much the same manner as synthetic rubber. PVAc and acrylic resins have good and excellent flexibility, respectively. All the films have good adhesion to bricks, plasterwork, cement, wood, and concrete. Polystyrene and styrene-butadiene have excellent adhesion to hydrophobic surfaces such as oil paint but poor adhesion to hydrophilic surfaces such as glass; the opposite is true of films based on PVAc and acrylic esters (Blackley 1997). All three films are washable, although styrene-butadiene paints gain their water resistance more slowly than PVAc or acrylic paints (*Field Applied Paints and Coatings* 1959: 47). PVAc paints are the most water sensitive of the three, and acrylic is by far the most water resistant. Styrene-butadiene and acrylic paints are completely resistant to hydrolysis, and the alkali resistance of PVAc resins is good as long as the binder does not contain polyvinyl alcohol.

Styrene-butadiene binders were most suited to interior applications, and although they were marketed for exterior paints, they were inferior to both PVAc and acrylics for this purpose. Furthermore, styrene-butadiene not only dried more slowly than alternative types but also was suited only to the formulation of white and pale tints due to its tendency to chalk over long exposure times, which spoiled the appearance of dark tints. However, pale tints tended to discolor on aging.

Styrene-butadiene is the slowest drying of the three: this is because the film dries by evaporation of the water, followed by coalescence of the polymer and oxidation of the butadiene. PVAc paints are faster drying and can be recoated after just two or three hours, while acrylics take just thirty minutes to dry and can be recoated in just ten minutes. PVAc paints have better exterior durability and color retention than styrene-butadiene and offer a more uniform, lower sheen. Acrylics are far superior to both styrene-butadiene and PVAc in terms of durability and ease of manufacture. Acrylics are more expensive but have better pigment-binding capacity, resistance to moisture, and color stability than the other types (Williams 1960: 155).

AVAILABILITY AND USE OF SYNTHETIC POLYMER EMULSIONS

Styrene-butadiene paints were introduced to the U.S. market in 1948, PVAc slightly later. PVAc paints were introduced to the British market in the late 1940s, although they do not appear in the trade press with any regularity until the early 1950s. PVAc remains the dominant binder for wall paints in Britain today. Although PVAc was initially more expensive than styrene-butadiene, the cost of the monomer gradually decreased in the United States throughout the 1950s, until by the end of the decade the two resins were similarly priced. Many U.S. manufacturers used PVAc resins in the formulation of exterior paints and primer sealers (*Field Applied Paints and Coatings* 1959: 47) but continued to use styrene-butadiene for top coats. However, throughout the 1950s PVAc resins increased in popularity in the United States, and their all-round good properties, coupled with their competitive price and ease of manufacture, meant that they captured a significant portion of the U.S. market by the late 1950s. Until the mid-1950s the U.S. market was dominated by homopolymer emulsion paints, but from that time on, the use of copolymers, typically PVAc and acrylic, increased considerably (Williams 1960: 33).

Although acrylic resins undoubtedly possessed superior characteristics in every respect, their expense meant that they were not widely used in either the United States or the United Kingdom on their introduction in the mid-1950s, and they were used only where superior durability was required. PVAc paints continued to dominate the British market, whereas styrene-butadiene was the preferred choice in the United States. By 1960 styrene-butadiene was still used in the largest volume for the production of wall paints, as the monomers were inexpensive and widely available, the emulsions had good storage stability, the resins were compatible with a wide range of modifiers, and a consistently uniform product was available; moreover, as it was the first binder introduced, there was a long product history (Williams 1960: 155). Today the U.S. interior wall paint market is dominated by vinyl acetate–acrylic copolymer. The best-quality exterior paints are 100% acrylic, and styrene-butadiene would be considered only if an extremely low price was required.

ACCEPTANCE AND USE OF SYNTHETIC RESINS

Despite their many good properties, paints based on synthetic resins were not always well received by the consumer, especially in the United Kingdom, whose decorators were notoriously conservative. Most British consumers remained mistrustful of paints based on synthetic resins until well after World War II, to the extent that until the early 1950s many manufacturers actively avoided advertising the components of their new paint binders (see plate 19). By contrast, those in the United States were so enamored with synthetic paints that when wartime shortages forced companies to revert to traditional materials, some postwar dealers fraudulently labeled their oleoresinous paints "plastic" and attempted to sell them at an inflated price ("The Plastics-Paint Racket" 1945: 98). Moreover, a feature of the U.S. trade press is the abundance of articles expounding new raw materials and describing how they should be formulated into paints. And whereas U.S. manufacturers proudly announced their latest developments, the British press showed a complete lack of interest in the chemical principles underpinning new materials. The peculiar disengagement displayed by both the professional and the trade press is difficult to fathom, and the degree to which this was the fault of the manufacturer or the painter is not clear: although the trade press regularly criticized the conservatism of British painters, the press itself appears to have done little to attempt to educate them in these matters.

A disinclination to be furnished with an understanding of the synthetic resins used in paint production seems to be a trait peculiar to the British paint industry. By contrast, the British plastics industry had recognized early on the value of a scientific forum to facilitate the exchange of ideas between industry, salesmen, and end users and in 1929 launched the journal *British Plastics* (Keller 1929: 4). However, the paint industry seems to have been content to continue to operate in ignorance, exemplified by one author, writing in 1932, who elucidated his understanding of the properties of fast-drying synthetic resins as their "special . . . automatic self-hardening properties" (Harrison 1932: 32–34).

The use of the word *synthetic* in fact appears to have been at the root of many of the problems: consumers were certainly suspicious of the word, as it tended to be associated with low-quality substitutes. Several authors lamented the fact that *synthetic* had been used to describe newly developed resins at all ("Synthetics" 1950: 41), as it was now very difficult to change that perception. One author asserted that manufacturers would not use the word in their advertising, resorting instead to subtler techniques such as relying on the brand name only. This is borne out in a study of the advertisements in the British press, few of which mention the components of the paints until well into the 1950s. This technique was reportedly successful, with manufacturers selling thousands of gallons of finishes containing synthetic resins without the buyer's knowledge (Snelling 1952a: 44). Indeed, one author commented, "Before the 1950s the idea of using a synthetic enamel was distasteful to many—the genuine article only would do. So one specified lead or hard gloss paints for wood and plaster and heat-resisting gloss paints for radiators, probably not realizing that the despised synthetic was the basis of many of these useful paints" (Mcfarlane 1963: 161).

An unfortunate side effect of this deception is that it is very difficult to ascertain which resins British manufacturers were using, or indeed the overall extent of use of coatings containing synthetic resins. There is no doubt that manufacturers were well aware of professional painters' mistrust of synthetic materials, as some went so far as to advertise the absence of synthetic resins as a feature of their paints (*The Decorator,* January 15, 1938: 35).

It is evident that this reluctance to fully engage with the chemical principles underlying paint formulation continued for a number of years, with manufacturers in both Britain and the United States relying on experimentation rather than rigorous scientific investigation to achieve success. The British Imperial Chemical Industries' somewhat haphazard approach to paint formulation is suggested by the company's statement that as late as 1947 alkyds were being formulated successfully in spite of an erroneous understanding of the polymerization processes taking place (Lilley 1959: 3). And as late as 1981 another author criticized the industry's inability to apply scientific principles to paint production, relying instead on empirical experimentation with materials that had been developed by the raw materials suppliers (Elliot 1981: 14).

The strength of resistance to synthetic paints by the British trade is certainly surprising, as evidenced by one author writing in 1942: "Articles have appeared in the daily press and even in some technical journals, hailing various substitutes as the only possible media for post-war use. It is probable that many of the authors of these articles are enthusiastic rather than experienced; possibly even thinly disguised journalists searching for something new, rather than practical men with knowledge of building materials" (Jay 1942: 171). By 1950, when the good properties of synthetic resins were no longer in question, prejudice against them remained strong: reportedly because "synthetic" was still equated with "inferior" and because alkyd resins in particular required more careful handling and a higher degree of skill in application than did traditional finishes ("Synthetics" 1950: 41).

Although the reasons for decorators' suspicion of new materials were varied, it must in part have been caused by the fact that throughout the twentieth century professional painters became increasingly divorced from the preparation of their materials. Within a very few years painters who had been accustomed to mixing their own paints from a limited range of materials were confronted with a variety of new products that would not tolerate the addition of a little extra turpentine, driers, or oil. As the number of increasingly sophisticated ready-mixed paints grew, the likelihood of painters using them inappropriately increased, and paint failure often resulted. Moreover, one of the first synthetic decorative paints introduced, nitrocellulose brushing lacquers, was replete with so many problems that it should come as no surprise that the reputation of paints based on synthetic resins was tarnished for many years to come.

It is evident that many shared the opinion that modern synthetic paints were inferior to those that professional painters mixed themselves, but implicit in this argument was the idea that painters knew their materials, whereas in reality they did not. The fact that their raw materials were natural substances meant that there was no reproducibility, and it was very difficult to ascertain precisely what the problem was if things went wrong. By contrast, paints based on synthetic resins were manufactured with batch-to-batch reproducibility, and most of the problems encountered with them were the result of unrealistic expectations of the paints themselves and their inappropriate use ("Familiar Criticism" 1954: 43). For example, many of the problems with synthetic paints reported to *The Decorator* throughout the 1930s stemmed overwhelmingly from such practices as applying them over old coatings or mixing them with incompatible materials. Some of the blame for the latter problem must lie with the manufacturer, who rarely printed the contents of the paints on the can.

The reluctance of manufacturers to label their paints resulted in some lively discussion between them and painters. Manufacturers claimed that there was little point in printing a list of ingredients that would be meaningless to painters, while painters complained that they would never be able to apply the paints correctly if they did not know what was in them (Bletchley 1954: 75). There were, however, more complex factors that discouraged manufacturers from listing their paints' ingredients: not only did it alert competitors to the paints' formulations, but it precluded manufacturers from altering the components if raw materials became unavailable or uneconomical. Moreover, in the inter- and postwar years in particular a declaration of the inclusion of a synthetic resin might harm sales.

Although the conservative nature of the professional painting industry in the United Kingdom played an important role in the poor reception of solvent-borne paints based on synthetic resins, the properties of the paints themselves were also to blame: early paints based on synthetic resins, in particular nitrocellulose lacquers and alkyds, suffered numerous problems: nitrocellulose lacquers dried too quickly to allow brushing out, and their solvent sensitivity made the application of a second coat impossible, whereas early alkyds suffered from poor stability in the can, sagging, poor brushing, and poor hiding power, to name but a few shortcomings. Moreover, both coatings were more expensive than traditional oleoresinous finishes.

The phenol-formaldehyde resins introduced in the late 1920s similarly received a mixed reception. Although excellent coatings were manufactured from modified and straight phenolics, the fact that they dried so quickly made users suspicious: before their introduction fast drying had been achieved by the addition of lots of resin or driers, both of which were detrimental to the film. Furthermore, because phenol-formaldehyde and alkyd resins were used as a direct replacement for natural gums and resins in oleoresinous finishes, the resulting paints were similar in both appearance and properties to traditional finishes. This enabled painters to make direct comparisons between the two types, and the shortcomings of many early synthetic resins quickly became apparent.

Oleoresinous finishes based on natural products thus remained popular for a number of years after synthetic alternatives were introduced: as long as there was a market for them, manufacturers continued to make them. Even high-grade oleoresinous paints were cheaper than alkyds, so many companies manufactured a lower-grade, cheaper line of paints based on traditional oils and resins alongside their more expensive alkyd range.[11]

The prewar U.S. market was certainly more receptive to the newly available synthetic resins than the British had been. This created some unlikely problems when their production stopped during the war. Synthetic resin finishes based on phenol-formaldehyde and oil-modified alkyds had enjoyed a brief success in the 1930s but became unavailable for domestic use during the war. This prompted some unscrupulous companies to buy up large quantities of canned paints from manufacturers, label them "plastic," and attempt to sell them to an unwitting public at an inflated price. The American National Paint, Varnish, and Lacquer Association was first alerted to this scam when it realized that no synthetic resins should be available for civilian use. In 1945 the association tested a number of these paints, finding a range of high-quality house paints through to standard or poor-quality finishes but none that contained synthetic resins ("The Plastics-Paint Racket" 1945). The contrast with the popularity of synthetic paints in the United Kingdom could not be greater.

Another important factor in the widespread acceptance of paints based on synthetic resins in the United States must also be its far larger retail market. Whereas the British paint industry did not attempt to enter the retail market with any determination until the early 1950s, U.S. companies had been advertising their products to this market sector for many years: Sherwin-Williams opened its first retail stores just outside Cleveland, Ohio, in 1911–14 (McDermott and Dyer 1991: 46), and DuPont was advertising paints for the retail market by the early 1920s. The launch of each new product was supported by massive advertising campaigns geared specifically to the home user: DuPont's "Brush" Duco, introduced in the mid-1920s, and household Dulux, introduced in the early 1930s, were certainly aimed squarely at this market.

By the early 1950s paints based on synthetic resins were becoming increasingly popular in the United Kingdom: references to synthetic resins certainly appear with increasing regularity in the British press in 1950 and 1951, and the usefulness of new finishes such as chlorinated rubber, emulsion paints, and synthetic enamels was further recognized when such finishes were used on the Festival of Britain buildings in 1951. The surfaces to be coated included asbestos, concrete, and plaster, difficult substrates on which traditional systems might have failed (Venison 1951).

Although oil-based finishes containing synthetic resins were initially ill received in the United Kingdom, when emulsion paints based on synthetic polymers were introduced in the late 1940s and early 1950s, they immediately became popular. This was because consumers were generally more willing to accept synthetic resins by that time but also because emulsion paints were a new type of product that filled perfectly a niche in the market for a material that was as easy to apply as water paint but as durable as oil paint. They had many other good properties, including easy cleanup, resistance to hydrolysis, and moisture permeability. Boosting their popularity was the dramatic rise in the do-it-yourself market in the early 1950s. These consumers were most concerned with easy application, low odor, and easy cleanup, and emulsion paints, with all these good properties, were an immediate hit. By the mid-1950s wartime austerity was easing, and sales of paint for decorative surfaces surged. By that time no special skill or experience was required to achieve a good paint job, and the excellent paint binders that are still in use today had become well established.

Chapter 7

The Conservation Implications of the Presence of Household Paints

The ubiquity of household paints means that conservators are increasingly being called on to preserve them. Although the paint industry has conducted extensive tests on the performance and durability of these coatings, the results seldom translate into useful data for the conservator, as the paint industry and the conservation profession approach the concept of durability from quite different perspectives.

Paints for household use are assessed by industry in terms of their ability to protect and decorate surfaces under a range of conditions, including extreme cycles of temperature and humidity, water resistance, exposure to UV radiation, contact with strong cleaning agents, and resistance to knocks, scuffs, and abrasion. Although a coating might not fail under these conditions, it is unlikely that its subtler characteristics such as gloss and color will remain unaltered. It is these properties that are often of greatest importance to the conservator.

Moreover, some early paint systems were designed to deteriorate gradually in order to retain an acceptable appearance: the controlled chalking of traditional oil-based exterior house paints might have kept the surface clean but in fact is a sign that the coating is slowly being eroded. Likewise for an exterior trim paint that was primarily decorative: it was preferable to choose a nonlightfast pigment that faded evenly rather than a lightfast one that caused streaks or blotches as it aged (Parker 1943: 322). Indeed, for household paints that are expected to be replaced every few years as fashions change or the coating becomes dirty, lightfast pigments and binders with good color retention might not be at the top of the manufacturer's list of priorities. For example, the most desirable properties of an interior wall paint are that it is easy to apply, has good coverage, is easy to clean up, is scrubbable, and is easily recoated. It might not matter that the gloss or color alters slightly over time, as long as this happens in an even manner.

Arguably one of the greatest difficulties for the conservator is establishing exactly what materials might be present in a household paint. This is especially the case for paints dating from the first half of the twentieth century, when synthetic resins and natural products were available simultaneously and wartime shortages resulted in great diversity in raw materials. Ascertaining what might be present in a paint is especially problematic with interior paints. Interior paints did not need to be as durable as exterior paints, so cost, drying speed, ease of application, and initial appearance usually took precedence over long-term durability: it did not really matter if an interior finish failed since it performed a primarily decorative function. But because durability was not necessarily a priority, a vast range of materials could be and were incorporated into them, which makes defining what a "typical" interior paint might comprise all but impossible. For example, a gloss or enamel paint could range from a high-grade binder of copal fused with refined linseed oil to cheap spirit-soluble coatings containing soft resins such as rosin or shellac.

The late nineteenth and early twentieth century saw the introduction of a number of new resins, including ester gum, phenolics, and alkyds. By the mid-twentieth century, in addition to linseed, a great number of oils were also available, including tung, soybean, oiticica, tall oil, safflower, dehydrated castor, and fish: these might be used alone or in admixture and could be fused with natural or synthetic resins. As the manufacturer's choice of raw material was dictated primarily by its cost and availability, it was common practice to substitute materials as prices fluctuated. Moreover, in much the same manner as artists' paint manufacturers produce a cheaper "student" grade of paint alongside their best products, commercial paint manufacturers produced different lines of paint of varying quality. This was apparent from the inception of the ready-mixed paint industry, when numerous companies even entered into competition with themselves, producing poor-quality paints under fictitious names and using their genuine company name only on their best products.

Although many paints were advertised for exterior or interior use, it was common for the manufacturer to put interior and exterior oil-based paints on different binders. For example, throughout the 1950s and 1960s, before synthetic resins had displaced natural products, manufacturers might base their less durable interior finish on natural products and reserve the more expensive synthetic resins for the durable exterior paints. Finally, there was no guarantee that the binder was the same for each color within a line of paints. Although it was necessary for white and pale tints to be based on a binder with good color and color retention, this did not matter for dark colors. It was therefore common to use a cheaper binder that might discolor on aging for dark tints: for this reason, painters were warned against mixing colors unless the manufacturer specifically stated that the range of paints was fully intermixable, as incompatible binders could ruin a paint if they were combined.

The conservation implications of all these factors are no doubt becoming clear. For those endeavoring to preserve a surface, especially if a coating has been chosen precisely for the finish it offers, any small changes in color, gloss, or texture can have a significant impact on appearance and interpretation. Furthermore, the solvent, aqueous, and mechanical cleaning systems used by conservators are all capable of altering these surface properties, and each type of finish has its own weaknesses. For example, conservators routinely alter the pH of their cleaning system to achieve optimum results, but the pH sensitivity of each system differs: tung is resistant to alkalis, whereas linseed is not, and alkyds have the least alkali resistance of all the oil-based coatings.

Conservators also need to be aware of the possible dangers of treatments involving moisture, heat, and adhesives. For example, all emulsion paints based on synthetic polymers are sensitive to aromatic hydrocarbon solvents, which are used as carrier solvents in certain adhesives and consolidants: therefore, care needs to be taken to choose an appropriate adhesive that will not alter the paint film. A paint's reaction to heat and moisture can also vary considerably. Paints based on animal glue binders remain both solvent and heat sensitive, and paints comprising soft, flexible polymers such as polyvinyl acetate and acrylic are softened considerably by heat. The moisture resistance of each binder likewise differs: all emulsion paints based on synthetic polymers are swelled by water, while the water resistance of alkyd resins is in fact inferior to good-quality oil and fossil resin binders.

It is not only conservation treatments that have the potential to alter the surface characteristics of household paints. Oil paints, distempers, and acrylics have a tendency to imbibe dirt; paints containing natural oils and resins yellow on aging, as do some oil-based paints kept in the shade; and the reflectance of gloss paints tends to decrease as the paint ages, especially if the paint contains a soft, friable resin. All these changes alter our perception and interpretation of the paint film and may make the film more vulnerable to conservation treatments.

The potential behavior of household paints in response to various conservation treatments is a little-studied area, and it is not possible to provide firm guidelines as to how each should be treated. It is possible, however, to discuss the strengths and weaknesses of each type and draw attention to potential problems.

OLEORESINOUS PAINTS BASED ON NATURAL, PHENOLIC, AND ALKYD RESINS

Although much has been written about the properties and durability of artists' oil paints, and their aging characteristics and reaction to conservation treatments are reasonably well researched and understood, the same cannot be said of household paints based on oil. Although commercial and artists' oil paints are broadly similar, in that they comprise pigment, oil, and driers, the paint manufacturer draws on a far broader range of raw materials. Whereas artists' oil paints are based on refined linseed oil, and to a lesser degree poppy and walnut, the paint manufacturer's might contain any number of oils, used alone or in admixture, and combinations of raw and treated oils are common. Varnishes might be added to a paint to modify properties such as gloss, hardness, and drying time, and the varnish itself could contain any number of the vast array of available raw ingredients and might be spirit or oleoresinous.

Furthermore, the increased gloss, hardness, and drying time conferred by the presence of a natural resin was always at the expense of the coating's durability and contributed to its discoloration, embrittlement, cracking, flaking, and eventual breakdown: the greater the quantity of natural resin, the shorter the life of the film. Commercial paints also contain numerous other additives, including extender pigments, biocides, and fungicides; their influence on the aging of the film is not understood. Furthermore, for a film that was not necessarily designed to last, cheaper, less durable pigments might be chosen. By contrast, the best-quality artists' oil paints are formulated with the best-quality, most durable materials.

There are also considerable differences in commercial paint formulation in different countries. For example, soybean oil was used much earlier in the United States than in Britain. Soybean had been incorporated into U.S. paints since it first became available around 1907, and by the mid-1930s it was being used in admixture with linseed and other fast-drying oils such as perilla and tung. By the early 1950s it was used extensively in the manufacture of oil-modified alkyds. However, the expense of soybean precluded its use in British paint production until the late 1950s or early 1960s, and even then it was used only in the formulation of pale-colored alkyd paints.

It is essential for anyone interested in preserving a surface to understand how many variables there are with regard to commercially available traditional oleoresinous paints and consequently how many areas of potential vulnerability there are. For example, spirit varnishes will remain solvent sensitive. Oils, linseed in particular, are prone to discoloration. The presence of natural resins almost certainly means that a coating will have discolored and become embrittled on aging. All oil-based coatings are susceptible to alkaline hydrolysis, and exposure to solvents can solubilize low-molecular-weight products and extract them from the film. The influence on aging of extender pigments and additives such as emulsifiers is also unclear.

Although generally considered more durable than coatings based on natural oils and resins, phenolic and alkyd resins too have their weaknesses and in fact might be less durable from a conservation perspective.

Phenolic Resins

Phenolic resins were used throughout the 1930s in weather- and corrosion-resistant paints and varnishes and are still used in the formulation of industrial finishes today. They are available as both modified and 100% types, and although both have excellent weather, water, acid, and alkali resistance, 100% types are superior to modified types in this respect. Due to the expense of 100% phenolics, they were rarely used in decorative coatings, which instead comprised phenolic modified with rosin or ester gum. The main shortcoming of phenolics from the conservator's perspective is their tendency to become brittle and to discolor. This discoloration derives from both the resin and the oil component and again depends on the amount and type present. Phenolics were dissolved in either tung or a mixture of tung and linseed: the discoloration of linseed is more pronounced than that of tung, and the amount of linseed present will influence the discoloration of the final film. The resin component of a phenolic resin also contributes to its discoloration: the use of intermediates such as cresol resulted in significant discoloration, as did the presence of rosin or ester gum, which was essential for rendering the phenolic soluble in oil.

Alkyd Resins

The properties of alkyd resins are dictated by the intermediates used in the manufacture of the resin and the amount and type of oil present. Although a great many acids and alcohols can be used, decorative paints are usually formulated with phthalic anhydride and glycerol or pentaerythritol. They are long oil, with an oil length between 60% and 80%. Short-oil alkyds have better color and gloss retention but have poor adhesion and are inflexible. Long-oil alkyds of the drying type have good adhesion and weathering properties, but their resistance to chemicals is relatively poor (Schweitzer 2006: 103). The polyester resin itself does not discolor significantly, so the discoloration is primarily dependent on the amount and type of oil used,[1] which might be linseed, soybean, safflower, perilla, or dehydrated castor. Paints based on pale-colored oils have better color and color retention. In much the same manner as oleoresinous binders, alkyds discolor if used indoors, as yellow chromophores are produced. Although these bleach in outside applications, the bleaching process results in the production of low-molecular-weight products that can migrate or be removed from the film.

All alkyds become very brittle on aging, and their extension to breaking point can reach 0%, meaning that they cannot flex at all without cracking. The interior durability of alkyds is good, but their exterior durability can only be described as fair. Although they are used in outdoor applications, their long-term exposure and their color and gloss retention are inferior to acrylics (Schweitzer 2006: 102) and to high-grade traditional oleoresinous finishes.

Light Stability of Oleoresinous Paints Based on Natural, Phenolic, and Alkyd Resins

The deterioration of natural resins and oils is brought about by exposure to oxygen and UV radiation. Although oxidation has the initially desirable effect of drying the paint film, its effect continues long after the film has hardened and contributes to its discoloration, embrittlement, and eventual breakdown. Certain pigments, for example, those containing copper, accelerate this process considerably (Ioakimoglou et al. 1999).

The light stability of traditional oleoresinous paints depends on two factors: the amount and type of oil present, and the amount and type of resin. All natural resins discolor and become brittle on exposure to light, and the greater the amount of resin, the greater this deterioration will be. As oil films age, low-molecular-weight products are produced: these can be removed from the film relatively easily by solvents. Semi-synthetic resins such as ester gum and maleic anhydride–reacted rosin have better color and color retention than natural resins, although they too discolor on aging. The discoloration of alkyds is dependent on the oil used, and paints having pale oils such as soybean and dehydrated castor offer the best color. Phenolic resins have poor initial color and poor color retention.

Water Resistance of Paints Based on Natural, Phenolic, and Alkyd Resins

All the oleoresinous binders discussed here are reasonably resistant to weathering; phenolics have the best water resistance; alkyds, the least. All will tolerate exposure to water in the form of cleaning, but only phenolics can be immersed. Oil paints and oleoresinous varnishes have greater water resistance and weathering properties than alkyds. Although both alkyds and oleoresinous binders are susceptible to alkaline hydrolysis, alkyds are far weaker in this respect. Indeed, they are frequently modified with other compounds to improve this property. Any wet cleaning of alkyds in particular therefore needs to proceed with caution, and the pH of any cleaning system needs to be carefully controlled.

Chemical Resistance of Paints Based on Natural, Phenolic, and Alkyd Resins

Oil paints and oleoresinous paints are resistant to weathering but have relatively poor chemical and alkali resistance, alkyds considerably more so than oleoresinous finishes (Schweitzer 2006: 108). Alkyds and oils have fair resistance to petroleum solvents and oils, but alkyd films are softened by alcohols and aromatic hydrocarbons and can be removed by ketones, esters, and chlorinated solvents. Fresh oil paints can also be removed by ketones, esters, and chlorinated solvents, although the solvent sensitivity of all films decreases as they age. The use of polar solvents can soften a film and remove the low-molecular-weight components from it. Solvents have the potential to disrupt the surface of a gloss paint, and the use of polar ones in particular should be undertaken with great care. Gloss surfaces can also be disrupted through exposure to alkaline reagents and even mechanical cleaning.

In an effort to increase the durability of alkyds, it was common to modify the molecule; one of the most widely used modifiers was styrene, which became available in the late 1940s. Styrenated oils to be used in conjunction with natural or synthetic resins were also introduced at the same time. Although the introduction of styrene improved the paint's drying time, hardness, and resistance to water and alkalis, the monomer's sensitivity to aromatic hydrocarbon solvents is conferred to any paint system that contains it. Although decorative coatings are rarely exposed to aromatic hydrocarbons, the solvents are commonly found in the conservator's range of materials, where they might be used as free solvents or as a solvent for certain adhesives.

The air-drying phenolics used in decorative paints and varnishes can be formulated to confer excellent resistance to moisture, alkalis, solvents, fresh water, deionized water, salt water, and mild acids (Schweitzer 2006: 187). They are undoubtedly the most chemically resistant of all the coatings discussed here.

NITROCELLULOSE LACQUERS

The formulation of nitrocellulose lacquers is extremely complex, and a vast array of resins, plasticizers, and solvents could be incorporated. The formulas could vary so much in terms of solids content, degree of nitration, and amount and quality of resins and plasticizers that it is very difficult to make generalizations.

Nitrocellulose lacquers are among the least durable of all the coatings discussed in this book: they scratch easily and have a tendency to crack as they age, and they discolor on exposure to UV radiation. They also remain solvent sensitive, a characteristic that is largely determined by the type of resin included. Early lacquers contained soft resins such as dewaxed dammar and shellac, which conferred their poor properties to the final film. Ester gum was another very cheap resin but was not suitable for high-grade lacquers due to its high solvent retention (where solvent remains in the dried film) and its tendency to produce a hazed film. Later lacquers incorporated synthetic oxidizing resins such as alkyds and phenolics, which increased the gloss, hardness, adhesion, and solvent resistance of the film considerably. The presence of alkaline pigments also promotes decomposition of the binder, although this weakness could be improved by the addition of an acid such as citric (Schweitzer 2006: 108).

Another area of major weakness in nitrocellulose lacquers is the high plasticizer content. The plasticizer remains mobile within the film and eventually leaches out, whereupon it either evaporates or is washed away if the surface is cleaned. Loss of plasticizer results in a weak, brittle film that may also undergo dimensional changes, with the resulting shrinkage causing cracking and flaking.

Light Stability of Nitrocellulose

The light stability of nitrocellulose is poor. In 1934 the U.S. National Research Council published the results of experiments conducted by the Hercules Powder Company into the effects of UV radiation on the discoloration of clear nitrocellulose lacquers. As the lacquers experimented on were those that were available when they were used in decorative paints, a discussion of these results is useful (West 1934: 308). Hercules discovered that the discoloration of clear lacquers was greater than one might expect from the individual ingredients; that the resin component caused more discoloration than the plasticizer; and that the larger the number of ingredients, the greater the discoloration, even though some of the individual ingredients were not light sensitive if used alone. There was clearly a synergistic effect operating. Hercules also found that some pigments extend the life of a film exposed to UV radiation.

Water Resistance of Nitrocellulose

Nitrocellulose dries to a hard, moisture-impermeable film. This makes it fully washable but can cause problems if moisture is trapped beneath, as blistering and flaking of the film can occur. The moisture permeability of nitrocellulose lacquers reportedly increased as the number of component ingredients increased (Schweitzer 2006: 108).

Chemical Resistance of Nitrocellulose

The chemical resistance of nitrocellulose lacquers is largely dependent on the resins incorporated. Broadly speaking, nitrocellulose has fair resistance to weak acids, poor resistance to strong acids, and poor resistance to weak and strong alkalis. The film is partly soluble in alcohols and soluble in ketones and esters but has good resistance to aromatic hydrocarbons and excellent resistance to aliphatic hydrocarbons. Lacquers with high dammar content have poor resistance to fats, waxes, and alcohols—a weakness that could be improved dramatically by the incorporation of a modified phenolic resin (Simmonds, Ellis, and Bigelow 1943: 447). The solvent sensitivity of dried nitrocellulose lacquers is a significant weakness, and careful testing is of paramount importance before any solvent-based products are put into contact with nitrocellulose.

TRADITIONAL WATER PAINTS: LIMEWASH, WHITEWASH, DISTEMPER, OIL EMULSIONS

Limewashes were used primarily to coat interior and exterior plaster and masonry. They are friable coatings that last only about a year unless modified with an oil, wax, or protein, which increases their life to about two or three years. Indeed, limewash is really a sacrificial coating in that it degrades more quickly than the substrate it protects. All whitewashes and distempers abrade and can be rubbed from a surface relatively easily. Distemper is a soft, porous coating that allows the passage of moisture through it, but this also means that it has a tendency to imbibe dirt, which cannot be washed off with water due to the moisture sensitivity of the film.

The introduction of casein binders improved the durability of water paints considerably, as once dry, casein becomes insoluble in water and solvents, although it is swelled by water. Casein is also sensitive to alkalis, so if it is washed with soap, swelling will occur, and if it is scrubbed in this condition, the film is easily damaged. Casein paints are also porous, meaning that they imbibe dirt. However, the dirt is not held on the film as firmly as happens with oil paints, so it is easier to wash away. Casein films were frequently washed with formaldehyde after application to increase the hardness of the film and to prevent the casein from swelling on subsequent exposure to moisture (Atwood 1934: 51).

Light Stability

The light stability of water paints is dependent on their consistent components. Water-based paints containing oils and resins, including oil-in-water emulsions, alkyd-in-water emulsions, and modified distempers and limewashes, are susceptible to the same oxidation and aging as the oleoresinous finishes described above, so are not discussed in any further detail here. Proteins such as animal glue and casein are relatively stable to oxidation and, within reasonable parameters of temperature and humidity, undergo little chemical change (Mills and White 1987: 89).

Water Resistance

Limewashes are moderately water resistant, and their resistance can be increased by the addition of waxes, oils, and proteins. A limewash coating will withstand only two to three years of weathering before it needs replacing.

The water resistance of distempers based on animal glue is poor. The binder does not oxidize, so it remains water soluble in the dried film. The binder is thus easily swelled, and dirt removal by aqueous means is impossible. The water resistance of casein coatings is fair, although casein is swelled if exposed to water.

Moisture is detrimental to proteins such as casein and animal glue: it encourages microbial and bacterial growth, which break down the protein, and also initiates hydrolysis of the peptide linkages in the molecule, which reduces the average molecular weight (Mills and White 1987: 89). These contribute to the deterioration of the binder, leaving it brittle and weakened.

Although oil or alkyd-in-water emulsion paints are considerably more durable than distempers and limewashes, it must be remembered that emulsified systems also contain quantities of water-soluble components, including protective colloids, emulsifiers, and soaps. These remain water soluble in the final film and are in danger of being solubilized if the film is exposed to water.

Chemical Resistance

The chemical resistance of water paints is also dependent on their constituent ingredients. The oils and varnishes that might be present in fortified distempers, limewashes, and oil emulsions are susceptible to attack from polar solvents and alkaline reagents in much the same manner as the oleoresinous finishes described above. Limewash is sensitive to acids; casein, to alkalis. Paint binders based on casein and animal glue are resistant to organic solvents.

SYNTHETIC POLYMER EMULSIONS

Emulsion paints based on styrene-butadiene, plasticized styrene, PVAc, and acrylic were a dramatic improvement on older water paints in terms of their durability, scrubbability, and resistance to water, but they too have inherent weaknesses. As a group, emulsion paints are swelled by water, ketones, esters, and aromatic solvents.

Light Stability of Emulsion Paints

The light stability of each system varies considerably: acrylics are the most stable; styrene-butadiene, the least. Acrylics are exceptionally stable to deterioration by oxygen and UV light. Even after prolonged exposure to UV radiation they show minimal chalking and little or no darkening (Schweitzer 2006: 100–102). Although also possessing excellent light stability, PVAc resins are inferior to acrylics in this respect. For example, methyl methacrylate requires a 2.5 times higher dose of radiation than does PVAc to achieve equal degradation (Warson 1972: 437).

Despite claims by manufacturers that plasticized styrene emulsions had good light stability, these paints did in fact discolor on aging (Graves 1952: 78). The discoloration of styrene derives from the oxidation of the unsaturated groups on the styrene monomer. The discoloration is exacerbated by the presence of sulfur, which was used as a distillation inhibitor in the manufacture of styrene, and also by the presence of free styrene monomer: the greater the quantity of free monomer, the more severe the discoloration (Matheson and Boyer 1952).

Discoloration is also a major flaw of styrene-butadiene systems and is more pronounced than that found in styrene systems. The discoloration derives from the styrene units and also from the fact that only one of the double bonds on the butadiene unit is saturated during polymerization, which leaves an unsaturated site in the molecule at which oxidation can occur, rendering the resin susceptible to the same types of aging and embrittlement found in rubbers. Until this property was improved, styrene-butadiene was used for interior applications only. Antioxidants such as phenyl-napthylamine and similar types and certain phenol-formaldehyde inhibitors were suggested as suitable additives to retard the embrittlement of these coatings (Warson 1972: 428–29).

Water Resistance of Emulsion Paints

The main weakness of all emulsion paints from a conservation perspective is their water sensitivity. Although all are marketed as scrubbable, the scrub resistance of each in fact varies considerably. Even if the film is not disrupted visually on scrubbing, all swell appreciably in the presence of water, and in this vulnerable state water-soluble surfactants and emulsifiers can be removed from the film, and mechanical damage is more likely. Acrylics have far superior water and scrub resistance than other systems, whereas PVAc resins have the least.

Although in the case of PVAc some of the water sensitivity originates in the polymer, most of the water sensitivity of all emulsion paints arises from the presence of water-soluble compounds. Warson (1972: 484–86) provides a detailed analysis of the water sensitivity of commercial paints based on synthetic polymer emulsions, and it is from this source that the following information is drawn.

The water sensitivity of a base film is not always dictated by the amount of emulsifier present but rather by whether the emulsifiers are solubilized in the film or exist as discrete particles. If solubilized, exposure to water might result in some swelling and lifting of the film, and if as separate particles, immediate blanching might be observed when in contact with water. However, as long as disintegration of the film is not the immediate effect, the removal of water-soluble components may actually increase the water resistance of the film in the long term. Polyelectrolytes such as sodium polymethacrylate or sodium carboxymethyl cellulose exhibit greater water sensitivity than nonionic materials such as hydroxyethyl cellulose or natural gum, as the ionized materials are leached from the film more easily. However, the presence of pigment, and the type of substrate to which the coating is applied, can alter significantly the moisture sensitivity. Finally, the action of rain on exterior coatings continually washes away water-soluble components, with the net result of increasing the water resistance of the film over time. With systems stabilized by surfactants only, rather than protective colloids or combinations of protective colloids and surfactants, the low molecular weight and mobility of the surfactant means that it becomes displaced from the surface of the pigment particles and migrates to the surface of the film: such a process might be observed in acrylic films and in PVAc resins formulated without protective colloids.

A physical manifestation of water sensitivity is water spotting, which frequently can be observed in PVAc coatings that have been exposed to condensation and is a significant flaw of PVAc paints. Although acrylics and styrene-butadiene systems might exhibit temporary water spotting, neither systems water-spot permanently (Williams 1960: 155).

Chemical Resistance of Emulsion Paints

Although wall and masonry paints are rarely exposed to harsh solvents, they are exposed to cleaning agents, and they are often required to coat alkaline surfaces. An important property, therefore, is their alkaline resistance. The alkaline resistance of all the binders varies considerably. Acrylics, styrene, and styrene-butadiene are almost completely resistant to alkaline hydrolysis, while all PVAc resins are alkali sensitive to some degree, which causes progressive water sensitivity and eventual disintegration of the film. The alkali resistance of a PVAc coating depends on its formulation: for example, the presence of PVAl reduces the water and alkali resistance of a PVAc coating considerably. But despite the fact that the alkali and water resistance of PVAc resins are inferior to other systems, it is sufficient to enable their use for outdoor exposure and in alkaline conditions if formulated appropriately.

Emulsion paints are also sensitive to organic solvents, and although they may never encounter such solvents in household applications, solvents form the basis of many conservation treatments. Although none are sensitive to aliphatic solvents, all are swelled and partially solubilized by alcohols, ketones, and aromatic hydrocarbons.[2]

SUMMARY

The range of materials used in the manufacture of household paints in the first half of the twentieth century was truly diverse. Although the properties of household paints have been well studied by industry, and the mechanisms by which they deteriorate and their expected life spans are well known, this knowledge is seldom capable of informing the conservator about appropriate conservation procedures. As discussed at the beginning of this chapter, the paint manufacturer and the conservator approach durability quite differently, and what might be considered an irrelevant change to the manufacturer could be disastrous for the conservator, as even small alterations in color, gloss, or texture might have significant ramifications for the interpretation or future study of artifacts.

As an area of research, the conservation of household paints has not received much attention, and it is therefore not possible to provide firm guidelines as to how such coatings should be preserved. But it is hoped that a discussion of the paints' chemical composition and the extent of their availability, periods of use, aging characteristics, and potential weaknesses will enable those charged with their care to make informed decisions about how to proceed.

Notes

Introduction

1. For a detailed discussion of the reasons for artists' choice of commercial materials, see Standeven 2004.
2. Unless otherwise stated, the information in this section originates from Lake, Ordonez, and Schilling 2004.
3. Analytical reports are held in Tate Gallery's Conservation Department. Paint was analyzed using THM GC-MS.
4. The analytical report is held in Tate Gallery's Conservation Department. Paint was analyzed using PyGC-MS.
5. The political interests underpinning Siqueiros's use of new materials is examined in detail in Hurlburt 1976 and *Portrait of a Decade* 1997.
6. Analysis gave a 98% match for PVA, and the fact that the medium was immiscible with the water-based ink it had been mixed with led the author to conclude that the PVA (or possibly PVB) was present in solution form. Mulholland 2010: 106.
7. It should be noted that "polymer tempera" paints bore no relation to traditional tempera painting, in which egg yolk is used as the binding media.
8. For an excellent discussion of the history of artists' casein paints, see Mulholland 2010.
9. Paula Dredge, conservator, Art Gallery NSW, Australia, personal communication, April 11, 2010. Confusing the matter further is the fact that the term *polymer paint* is more frequently used to describe waterborne paints than oil-based products.

Chapter 1

1. Book review of Maximilian Toch, *The Chemistry and Technology of Paints,* 3rd ed. (London, 1926), in *The Decorator*, April 22, 1926, 773.
2. "Chalking" describes the loose powdery deposit left on the surface of a paint film as it ages and deteriorates, although it indicates that the coating is eroding. Controlled chalking was used to retain a clean, bright surface, especially for exterior paints. Today's coatings are not designed to chalk.
3. For a comprehensive introduction to twentieth-century pigments, see De Keijzer 1989.

Chapter 2

1. For a detailed description of the drying mechanism of oils, see, e.g., Bentley and Turner 1997.
2. Crystols were manufactured by the Archer-Daniels-Midland Company in the United States.
3. Advertisements appeared in *American Paint Journal*, June 27, 1949, 28–29; July 25, 1949, 70–71.

4 Livering occurred from a reaction between the pigment and binder, which caused the paint to thicken irreversibly to a rubberlike consistency.

5 G. D. Patterson, Monthly Resume Report, February 1928, acc. 2247, Series II, box 10, DuPont Archive, Hagley Museum and Library, Wilmington, DE.

6 For a detailed discussion of these industrial coatings, see, e.g., Stoye, Freitag, and Beuschel 1996.

7 The "rubber" referred to by Wolf is stryene-butadiene copolymer.

Chapter 3

1 Advertisement from Reardon Company, *National Painters Magazine*, September 6, 1936; rpt. Sutermeister and Browne 1939: 329.

2 "National Historic Chemical Landmarks," *American Chemical Society*, http://acswebcontent.acs.org/landmarks/landmarks/kem/kem_paint.html. Accessed September 17, 2009.

Chapter 4

1 Public Affairs Department Records, n.d., acc. 1410, box 46, 1, DuPont Archive, Hagley Museum and Library, p. 1.

Chapter 5

1 "Wood oil" or "China wood oil" were alternate names for tung oil.

2 Arthur Hopgood, interview by author, December 5, 2001; Ken Arbuckle, interview by author, March 13, 2002. Former employees of Lewis Berger & Sons.

3 Tom Graham, letter to author, May 16, 2002.

4 The work was carried out by M. J. Callahan at the Pittsfield works and by W. C. Arsem, E. S. Dawson, and K. B. Howell at the Schenectady laboratory of General Electric.

5 G. D. Patterson, Experimental Station Project Files: 1927–37, acc. 2247, Series I, boxes 7, 9; Series II, boxes 10, 11, DPA, HML.

6 Patterson, Monthly Resume Report, February 1928, acc. 2247, Series II, box 10, DuPont Archive, Hagley Museum and Library, Wilmington, DE.

7 Lithopone is a white pigment that comprises a mixture of barium sulfate and zinc sulfide. It is more suited for interior paints.

8 *United States vs. ICI: Chronological Outline*, n.d., DuPont Archive, Hagley Museum and Library, Wilmington, DE, n.p.

9 Brian Gilliam, former ICI employee, interview by author, October 12, 2000.

10 The information in this section has been taken from Cheetham and Pearce 1943.

11 Ken Arbuckle, interview by author, March 13, 2002.

12 Arthur Hopgood, interview by author, December 5, 2001.

13 Barbara Meyer, email to author, February 12, 2001.

Chapter 6

1 BON red is a metallized azo pigment based on either calcium or manganese salts.

2 The U.S. company Dewey & Almy Chemical was among the first to introduce an internally plasticized resin in the United States in 1954, Everflex G. (*American Paint Journal*, May 17, 1954, 94).

3 Wollastanite is a calcium inosilicate mineral (CaSiO3) that may contain small amounts of iron, magnesium, and manganese substituting for calcium.

4 PVAl is produced via the partial hydrolysis of PVAc.

5 Rohm & Haas, German Patent 295,340 (June 5, 1915).

[6] British Sign & Graphics Association: Industry News, www.bsga.co.uk/news_stories/news_item78_perspex.html. Accessed August 1, 2009.

[7] Ibid.

[8] The first advertisement for Rhoplex AC-33 appeared in the *American Paint Journal*, February 23, 1953, 17–20.

[9] Although acrylic is easily cleaned from a manufacturer's perspective, the removal of imbibed dirt from acrylic paintings is a significant problem in paintings conservation.

[10] Borden, Dow, General Tire, Goodrich, Monsanto, Morton Chemical, National Starch, Polyvinyl Chemicals, Reichold, Rohm & Haas, and UBS.

[11] Graham North, former Cray Valley physician, interview by author, June 12, 2002.

Chapter 7

[1] Ray Legetter, Mike Williams, David Elliot, interviewed by Tom Learner, Fotini Koussiaki, Fransesca Capitelli, ICI Paints, Slough, March 12, 1999. Tate Archive.

[2] Tate London has conducted considerable research into the cleaning of acrylic paintings, which includes a study of the swelling and leaching of materials from acrylic paints. www.tate.org.uk/research/tateresearch/majorprojects/conservation_modernpaints.htm. Accessed November 10, 2009.

References

Adair, J. 1938. Flat wall finishes. *The Decorator,* January 15, 55–57.

Alkyd resins. 1946. *The Decorator,* February 15, 52.

American Chemical Society. National historic chemical landmarks. http://acswebcontent.acs.org/landmarks/landmarks/kem/kem_paint.html. Accessed September 17, 2009.

Applying synthetic resin finishes. 1949. *The Decorator,* July 15, 85.

Armitage, F. 1955. A history of paints and varnishes in Great Britain, 6: A period of expansion. *Paint Manufacture,* June, 229.

Atwood, F. C. 1934. Modern casein paints. *The Decorator,* May 15, 46–48.

Barron, H. 1945. *Modern Plastics.* London: Chapman & Hall.

Beavers, E. M. 1949. Alkyd resin technology. *American Paint Journal,* April 11, 38–46.

Bentley, J., and G. P. A. Turner. 1997. *Introduction to Paint Chemistry and Principles of Paint Technology.* 4th ed. London: CRC Press.

Bevan, E. A., and R. S. Robinson. 1951. Alkyd resins. In *Synthetic Resins and Allied Plastics,* ed. R. S. Morrell. Oxford: Oxford University Press.

Blackley, D. C. 1997. *Polymer Latices: Application of Latices.* 2nd ed. London: Chapman & Hall.

Bletchley, R. L. 1954. Ingredients of ready-mixed paints. Letters to the editor. *The Decorator,* February 15, 75.

Bodied oil in exterior house paints. 1945. *The Decorator,* August 15, 28–31.

Bohannon. 1922. Present-day methods of varnish manufacture. *The Decorator,* January 22, 237.

Borchardt-Hume, A., ed. 2008. *Rothko.* London: Tate Publishing.

Borders, A. M., R. D. Juve, and L. D. Hess. 1946. Styrene-diene resins in rubber compounding. *Industrial and Engineering Chemistry* 38 (9): 955–58.

Boxall, J. 1978a. A history of paint technology, part two: 18th century to mid-19th century. *Paint Manufacture,* June, 23–34.

______. 1978b. A history of paint technology, part four: The 20th century. *Paint Manufacture* (September): 18–23.

Bradley, T. F. 1951. Nonedible soybean oil products. In *Soybeans and Soybean Products,* vol. 2, ed. K. S. Markely. New York: Interscience.

Bragdon, C. R. 1954. Glycerine alkyds tailored to need. *American Paint Journal,* March 29, 76–86.

Breek, R., and W. Froentjes. 1975. Application of pyrolysis gas chromatography on some of Van Meegeren's faked Vermeers and Pieter de Hooghs. *Studies in Conservation* 20: 183–89.

British Sign & Graphics Association. Industry news. www.bsga.co.uk/news_stories/news_item78_perspex.html. Accessed August 1, 2009.

Browne, F. L. 1941. Two-coat system of house painting. *Industrial and Engineering Chemistry* 33 (7): 900–910.

Brushable lacquers. 1925. *The Decorator,* August 22.

Brushing Belco. 1928. *The Decorator,* December 22, 456–59.

Brushing cellulose. 1938. *The Decorator,* April 15, 85, 111.

Brushing cellulose lacquers. 1930. *The Decorator,* April 22, 10.

Brushwell, W. 1954. Tall oil: Its function in the paint industry, parts I and II. *American Paint Journal,* August 16, 72–96; September 6, 60–84.

The building exhibition. 1949. *The Decorator,* December 15, 90–94.

Burr Price. 1943. Paints: Present and postwar. *Architectural Record,* June, 81.

Carlyle, L., et al. 2008. The substance of things. In *Rothko,* ed. A. Borchardt-Hume. London: Tate Publishing.

Cayley, B. 1926. Lecture on paint and painting. *The Decorator,* May 22, 844–46.

———. 1927. Modern paint and painting. *The Decorator,* April 22, 744–49.

Cellulose as an interior wall finish. 1934. *The Decorator,* July 16, 32.

Cellulose for new plaster. 1946. *The Decorator,* March 15, 60.

Cellulose paints: An interesting conference. 1928. *The Decorator,* November 22, 376–82.

Characteristics of boiled oil. 1925. *The Decorator,* May 22, 908.

Chatfield, H. W. 1953. *Varnish Constituents.* London: L. Hill.

Cheetham, H. C., and W. T. Pearce. 1943. Alkyd emulsion paints. In *Protective and Decorative Coatings,* vol. 3., ed. J. Mattiello. New York: John Wiley & Sons; London: Chapman & Hall.

Christensen, L. 1950. Safflower oil. *American Paint Journal,* February 6, 54–63.

Coffignier, C. 1922. Enamel paints. *The Decorator,* September 22, 80.

Coker, J. N. 1957. Poly(vinyl alcohol) as an emulsifying agent in vinyl polymerizations. *Industrial and Engineering Chemistry* 49 (3): 382–85.

Composition of distempers. 1933. *The Decorator,* July 15, 67.

Cook County Paint & Wallpaper Assn. hears about alkyd and emulsion paints. 1954. *American Paint Journal,* February 22, 74–75.

Coolidge, C. 1929. Monthly Summary Report, August. In G. D. Patterson: Experimental Station Project Files, acc: 2247, Series II, box 10. DuPont archive, Hagley Museum and Library, Wilmington, DE.

———. 1936. Letter to John Marshall, November 6. Experimental Station Project Files, Box 11, DuPont Archive, Hagley Museum and Library, Wilmington, DE.

Crook, J., and T. Learner. 1999. *The Impact of Modern Paints.* London: Tate Publishing.

Cross, M. 2008. The complexities of *Woman in a Courtyard* (1933) by Julian Trevelyan. In *Proceedings of the Modern Paints Uncovered Symposium, May 16–19, 2006, Tate Modern, London,* ed. T. J. S. Learner, P. Smithen, J. W. Krueger, and M. R. Schilling. Los Angeles: Getty Conservation Institute.

Cruickshank Smith, J. 1924. *The Manufacture of Paint* (London, 1924). Book review. *The Decorator,* September 22, 231–32.

———. 1928. Cellulose lacquers: Their appeal to the painter. *The Decorator,* October 22, 268–70.

Damitz, F. M., J. A. Murphy, and J. J. Mattiello. 1943. Varnishes. In *Protective and Decorative Coatings,* vol. 3, ed. J. Mattiello. New York: John Wiley & Sons; London: Chapman & Hall.

Davis, R. J. 1954. Polyvinyl acetate as a paint vehicle. *American Paint Journal,* May 24, 78–82.

De, S. K., and J. R. White, eds. 2001. *Rubber Technologist's Handbook.* Shrewsbury, U.K.: Smithers Rapra.

De Keijzer, M. 1989. The colourful twentieth century. In *Modern Art: The Restoration and Techniques of Modern Paper and Paints: Proceedings of a Conference Jointly Organized by UKIC and the Museum of London, May 22, 1989.* London: Institute for Conservation.

Dispersed rubber. 1931. *Synthetic and Applied Finishes* (May): 18, 23.

Distempers and waterpaints. 1938. *The Decorator,* February 15, 51.

Dulac, R. 2007. *Industrial Cold Adhesives: A Practical Handbook for the Maker and User.* London: Frazer Press.

DuPont. 1949. *Refinisher News,* July. [Internal newsletter.]

Earhart, K. A. 1954. Trends and factors affecting fatty acid usage in alkyd resins. *American Paint Journal,* March 22, 74–90.

Elliott, W. T. 1981. Paint technology in the 80s: Evolution or revolution. *Australian OCCA Proceedings & News* (November): 11–14.

Enamels and enamelling. 1923. *The Decorator,* July 23, 53.

Erbil, H. Y. 2000. *Vinyl Acetate Emulsion Polymerization and Copolymerization with Acrylic Monomers.* Boca Raton, FL: CRC Press.

Familiar criticism. 1954. *The Decorator,* January 15, 43.

Fasig, E. W. 1926. Rosin for varnishes. *The Decorator,* July 22, 54.

Feldman, D., and A. Barbalata. 1996. *Synthetic Polymers: Technology, Properties, Applications.* London: Chapman & Hall.

Field Applied Paints and Coatings. 1959. Washington, DC: Building Research Institute.

Flat oil finishes. 1957. *The Decorator,* July 15, 43–44.

Flat paints, flat wall paints and flat enamels. 1931. *The Decorator,* March 23, 663.

Flatting oils. 1932. *The Decorator,* April 22, 6.

Fonrobert, E., et al. 1930. Albertol copals as an aid to modern varnish making processes. *Synthetic and Applied Finishes* 1 (6): 88.

Forum on latex paints attracts record crowd at New York Production Club. 1953. *American Paint Journal,* March 16, 9–11.

Fowler, D. E., and C. B. Westerhoff. 1952. The development of elastomer latices for emulsion paints. *American Paint Journal,* November 24, 74–90.

Future of wartime paint. 1948. *The Decorator,* January 15, 44.

Gardner, H. A. 1935. Looking into the future: Rubber paints. *The Decorator,* May 15, 49–52.

Gardziella, A., L. Pilato, and A. Knop. 1999. *Phenolic Resins: Chemistry, Applications, Standardization, Safety, and Ecology.* New York: Springer.

"Gelled" paint: New United States development. 1955. *The Decorator,* January 15, 69.

Gettens, R. L., and G. L. Stout. 1966. *Painting Materials.* New York: Dover.

Grades of drying oils. 1924. *The Decorator,* July 22, 75–76.

Graves, W. L. 1952. Polystyrene emulsions for coatings, part 1. *American Paint Journal,* December 22, 78–85.

———. 1953. Polystyrene emulsions for coatings, part 2. *American Paint Journal,* January 5, 60–70.

Green, R. B. 1953. Externally plasticized latices in paints. *Industrial and Engineering Chemistry* 45 (4): 726–28.

Gutiérrez, J. L. 1956. *From Fresco to Plastics.* Ottawa: National Gallery of Canada.

Haines, H. W., Jr., D. Griffin, and A. J. Da Valle. 1954. Staff-industry collaborative report: Resin and paint production, 1954 style. *Industrial and Engineering Chemistry* 46 (10): 2010–22.

Harrison, A. W. C. 1931. The application of flat wall paints. *The Decorator,* June 22, 862–64, 901–2.

———. 1932. Modern ultra-quick drying finishes. *The Decorator,* June 22, 32–34.

Heckel, G. B. 1928. *The Paint Industry: Reminiscences and Comments.* St. Louis: American Paint Journal Co.

Higher methacrylates: Chlorinated polymers. 1960. ICI, General Chemicals Division: Development Department, Development note 17444 (November). ICI Archive, Catalyst, GCD 63/1/6179.

Hochheiser, S. 1986. *Rohm and Haas: History of a Chemical Company.* Philadelphia: University of Pennsylvania Press.

Holmberg, K. 1987. *High Solids Alkyd Resins.* New York: Marcel Dekker.

Horn, M. B. 1960. *Acrylic Resins.* New York: Rheinhold.

Hounshell, D. A., and J. K. Smith. 1988. *Science and Corporate Strategy: DuPont R&D, 1902–1980.* New York: Cambridge University Press.

Hurlburt, L. P. 1976. The Siqueiros Experimental Workshop: New York, 1936. *Art Journal* (Spring): 237–46.

Husen, of Krumbhaar Chemicals, addresses New York vehicle manufacturers. 1952. *American Paint Journal,* October 20, 9–11.

Ianni, J. D. D., L. D. Hess, and W. C. Mast. 1951. Butadiene-styrene resinous copolymers. *Industrial and Engineering Chemistry* 43 (2): 319–24.

Ideal home exhibition. 1928. *The Decorator,* March 22, 674.

Institute Brains Trust. 1952. *The Decorator,* May 15, 61–67.

Ioakimoglou, E., S. Boyatzis, P. Argitis, A. Fostiridou, K. Papapanagiotou, and N. Yannovits. 1999. Thin-film study on the oxidation of linseed oil in the presence of selected copper pigments. *Chemistry of Materials* 11 (8): 2013–22.

Jay, B. A., comp. 1942. Materials: The post war outlook. *Architect & Building News,* June 26: 171–81.

Jennings, A. S. 1925. Modern varnishes and varnishing. Review. *The Decorator,* March 23, 723.

Jensen & Nicholson Ltd. 1930. Paint and its part in architecture. Publications received. *The Decorator,* October 22, 271.

Jolly, V. G. 1930. Oil varnishes: Their manufacture, properties, and defects. *The Decorator,* January 22, 522–26.

Jordan, L. A. 1929. Tung oil: With special reference to the possibilites of production within the British Empire. *The Decorator,* July 22, 58–60.

Keller, A. V. 1929. Synthetic resins: Their history, moulding technique, and commercial application. *British Plastics* (June): 4.

Kelly, A. A. 1927. Hints on prepared paint. *The Decorator,* May 23, 794–95.

Khandekar, N., et al. 2008. The re-restoration of Donald Judd's *Untitled* (1965)." In *Proceedings of the Modern Paints Uncovered Symposium, May 16–19, 2006, Tate Modern, London,* ed. T. J. S. Learner, P. Smithen, J. W. Krueger, and M. R. Schilling. Los Angeles: Getty Conservation Institute.

Kienle, R. H. 1949. Alkyd resins: Development of and contributions to polymer theory. *Industrial and Engineering Chemistry* 41 (4): 726–29.

Kienle, R. H., and C. S. Ferguson. 1929. Alkyd resins as film-forming materials. *Industrial and Engineering Chemistry* 21 (4): 349–52.

Lake, S., E. Ordonez, and M. Schilling. 2004. A technical investigation of paints used by Jackson Pollock in his drip or poured paintings. In *Modern Art, New Museums, Contributions to the Bilbao Conference, 13–17 September 2004,* ed. A. Roy and P. Smith, 137–41. London: International Institute for Conservation of Historic and Artistic Works.

Langton, H. M. 1943. General introduction. In *Synthetic Resins and Allied Plastics,* 2nd ed., ed. R. S. Morrell. Oxford: Oxford University Press.

Latex finishes; coating compositions and the like. 1930. *Surface and Applied Finishes* (August): 76.

Learner, T. J. S. 2008. Modern paints: Uncovering the choices. In *Proceedings of the Modern Paints Uncovered Symposium, May 16–19, 2006, Tate Modern, London,* ed. T. J. S. Learner, P. Smithen, J. W. Krueger, and M. R. Schilling. Los Angeles: Getty Conservation Institute.

Leclerc, D., and M. Barclay. 1992. *The Crisis of Abstraction in Canada: The 1950s.* Ottawa: National Gallery of Canada.

Lewin, M., and E. M. Pearce. 1998. *Handbook of Fiber Chemistry.* New York: Marcel Dekker.

Lilley, H. S. 1943. Phenolic resins in oil, varnish and enamel manufacture. In *Synthetic Resins and Allied Plastics,* 2nd ed., ed. R. S. Morrell. Oxford: Oxford University Press. [3rd ed., 1951.]

———. 1959. Development of postwar 'Dulux.' July 10. Transcript.

Limewash coatings. 1947. *The Decorator,* September 15, 36–38.

Limewash coatings. 1956. *The Decorator,* July 15, 53–56.

Linseed oil. 1949. *The Decorator,* February 15, 53–54.

Lodge, R. G. 1988. A history of synthetic painting media with special reference to commercial materials. In *Preprints of Papers Presented at the 16th Annual Meeting, New Orleans, June 1–5,* 118–27. Washington, DC: American Institute for Conservation.

Mann, L. C. 1952. The markets: Forecast for 1952. *American Paint Journal,* January 7, 41.

The markets. 1952. *American Paint Journal,* December 1, 35.

The markets. 1954. *American Paint Journal,* April 19, 50.

The markets: Dry colors. 1952. *American Paint Journal,* November 24, 49.

The markets: Miscellaneous materials. 1952a. *American Paint Journal,* June 2, 41.

The markets: Miscellaneous materials. 1952b. *American Paint Journal,* June 23, 39.

The markets: Miscellaneous materials. 1953. *American Paint Journal,* February 9, 47–49.

The markets: Varnish gums, resins. 1952. *American Paint Journal,* July 7, 49.

Martens, C. R. 1961. *Alkyd Resins.* New York: Reinhold.

Martin, S. R. W. 1947. Synthetic resins in the paint and varnish industry. *Plastics Monograph,* no. 15.

Matheson, L. A., and R. F. Boyer. 1952. Light stability of polystyrene and polyvinylchloride. *Industrial Engineering Chemistry* 44 (4): 867–74.

Matthieson, A. L. 1906. Varnish and its application. *The Decorator,* May 22, 320–22.

McDermott, K., and D. Dyer. 1991. *America's Paint Company: A History of Sherwin-Williams.* Cambridge, MA: Winthrop Group.

Mcfarlane, A. A. 1963. Introduction. Paints: general SfB V. A&BN Feature-13. *Architect & Building News,* January 30, 161.

Mills, J., and R. White. 1994. *The Organic Chemistry of Museum Objects.* 2nd ed. Oxford: Butterworth-Heinemann.

Molloy, E., and P. Reilly, eds. 1948. *British Catalogue of Plastics: A Plastics Encyclopaedia.* London: National Trade Press.

Morris, H. N. 1931. Colour and its industrial uses: The new rubber-cellulose medium. *Surface and Applied Finishes* (January): 138–39.

Morton, M., ed. 1987. *Rubber Technology.* 3rd ed. New York: Van Nostrand Reinhold.

Mossman, S. T. I., and P. J. T. Morris, eds. 1994. *The Development of Plastics.* Cambridge: Royal Society of Chemistry.

Mulholland, R. 2010. From dreams and visions and things not known: Technique and process in David Smith's drawings. Ph.D. thesis, Royal College of Art.

Munns, E. A. 1932. Quick-drying products. *The Decorator,* September 15, 46.

Naidus, H. 1953. Emulsion polymers for paint. *Industrial Engineering Chemistry* 45 (4): 712–17.

A new method of house decoration. 1927. *The Decorator,* April 22, 740.

New products and equipment. 1952. *American Paint Journal,* October 27, 92.

Nitro-cellulose. 1926. *The Decorator,* August 23, 117–18.

Nobel Chemical Finishes Ltd. 1931. *Synthetic and Applied Finishes* (May): 23.

Notes from manufacturers: Styrene Copolymers Ltd. 1948. *The Decorator,* August 15, 99.

Odorless solvents and filtration discussed by New York vehicle manufacturers. 1952. *American Paint Journal,* November 24, 30.

Oiticica oil. 1942. *The Decorator,* November 15, 17–18.

Olagoke, O. 1997. *The Development of Plastics.* New York: Marcel Dekker.

Oliver, C. E. 1928. The future of brushing lacquers. *The Decorator,* April 23, 747–48.

Parker, D. H. 1943. Exterior trim paints. In *Protective and Decorative Coatings,* vol. 3, ed. J. Mattiello. New York: John Wiley & Sons; London: Chapman & Hall.

Patterson, G .D. Experimental Station Project Files: 1927–37, Acc. 2247, Series I, Boxes 7, 9; Series II, Boxes 10, 11, DuPont Archive, Hagley Museum and Library, Wilmington, DE.

The plastics-paint racket. 1945. *Modern Plastics* (November): 98.

Plastic emulsion paints. 1952. *The Decorator,* January 15, 52.

Portrait of a Decade, 1930–1940: David Alfaro Siqueiros. 1997. Mexico City: Instituto Nacional de Bellas Artes.

Powers, P. O. 1951. Rosin-modified phenolic resins. *Industrial Engineering Chemistry* 43 (8): 1770–74.

Prane, J. W. 1954. The use of safflower oil in paint vehicles. *American Paint Journal,* March 15, 76–92.

Public Affairs Department Records. n.d. Acc. No. 1410, Box 46, 1, DuPont Archive, Hagley Museum and Library, Wilmington, DE, p. 1.

Radcliffe, R. S. 1943. Casein paints. In *Protective and Decorative Coatings,* vol. 3, ed. J. Mattiello. New York: John Wiley & Sons; London: Chapman & Hall.

Raffé, W. G. 1928. Patent medicine and patent paints. *The Decorator,* September 22, 219.

Raw and boiled oil. 1937. *The Decorator,* August 16, 80.

Reboul, P. 1994. Britain and the Bakelite revolution. In *The Development of Plastics,* ed. S. T. I. Mossman and P. J. T. Morris. Cambridge: Royal Society of Chemistry.

Richter, W. M. 1929. Letter to W. A. Hart, February 22. William P. Allen Papers, Acc. No. 1832, DuPont Archive, Hagley Museum and Library, Wilmington, DE.

Rivenc, R., J. Barten, F. Esmay, T. Learner, and C. Stringari. 2010. Simply White: The diverse painting materials of Robert Ryman. In *Postprints of the Paintings Speciality Group meeting, American Institute for Conservation Conference 2009, Los Angeles,* ed. B. Buckley. Washington, DC: American Institute for Conservation.

Rohm & Haas announces new acrylic monomer plant. 1953. *American Paint Journal,* March 16, 44.

Rohm & Haas Co. 1953. *Synthetic Resins for Coatings.* Philadelphia: University of Pennsylvania Press.

Ronal, K. S., and S. M. Weisberg. 1954. Modified proteins for stabilizing latex paints. *Industrial Engineering Chemistry* 46 (4): 774–77.

Rubber in paints and varnishes. 1942. *The Decorator,* January 15, 29.

Rubber paints. 1938. *The Decorator,* July 15, 68.

Rubber paints. 1954. *The Decorator,* January 15, 44–45.

Ryden, L. L., N. G. Britt, and R. D. Visger. 1949. Emulsion paints from a butadiene-styrene emulsion copolymer. *American Paint Journal,* October 17, 74.

Sanderson, J. McE. 1934. Rosin derivatives in paint products. *Industrial Engineering Chemistry* 26 (7): 711–15.

Scheffele, W. F., and N. G. Tompkins. 1959. Formulation of water-thinnable paints from emulsifier-free vinyl acetate copolymer dispersions. *Industrial Engineering Chemistry* 51 (2): 114.

Scholl, E. C. 1952a. Latex paint: The formulation and manufacture, part 1. *American Paint Journal,* August 25, 64–72.

______. 1952b. Latex paint: The formulation and manufacture, part 2. *American Paint Journal,* September 8, 60–74.

______. 1952c. Latex paint: The formulation and manufacture, part 3. *American Paint Journal,* September 29, 66–76.

Scholz, H. A. 1953. History of water-thinned paints. *Industrial Engineering Chemistry* 45 (4): 710–11.

Schroeder, B. W. 1952. New York vehicle manufacturers discuss uses of fatty acids. *American Paint Journal,* January 28, 57–58.

Schultz, M. J. 1969. How to choose one of today's "wonder paints." *Popular Mechanics* 131 (5): 196–98, 228.

Schweitzer, P. A. 2006. *Paint and Coatings: Applications and Corrosion Resistance.* Boca Raton, FL: Taylor & Francis.

A short history of ICI Paints Division. 1979. *Polymers Paint & Colour Journal,* May 30, 30.

Should high-class decorators use ready mixed paints? 1922. *The Decorator,* July 22, 14–15.

Simmonds, H. R., C. Ellis, and C. W. S. Bigelow. 1943. *Handbook of Plastics.* London: Chapman & Hall.

Singer, E. 1954a. Fundamentals of paint varnish and lacquer technology, Chapter X: The clear oleoresinous finish (Part III). *American Paint Journal,* May 17, 78–92.

———. 1954b. Fundamentals of paint varnish and lacquer technology, Chapter XII: Exterior finishes for houses and buildings. *American Paint Journal,* August 2, 58–77.

Siqueiros, D. A. 1975. Towards a transformation of the plastic arts (New York, 1934) [expansion of his 1924 "declaration"]. In D. A. Siqueiros, *Art and Revolution.* Trans. Sylvia Calle. London: Lawrence and Wishart.

Snelling, J. 1952a. Facts and theories on painting XI. *The Decorator,* February 15, 50–52.

———. 1952b. Know your paint II. *The Decorator,* April 5, 59–60.

———. 1952c. Know your paint: IV. *The Decorator,* June 15, 61–62.

———. 1953a. Know your paint XI. *The Decorator,* January 15, 46–48.

———. 1953b. Know your paint XIII. *The Decorator,* May 15, 52–55.

Snow, C. I. Dulux finishes. n.d. Sent under covering letter to J. Ferguson. [Copy in author's possession.]

Some points in the use of enamel paints. 1905. *The Decorator,* August 22, 29–30.

Soybean oil. 1936. *The Decorator,* May 15, 47.

Spiller, G. 1952. Manufacture and use of pentaerythritol. *American Paint Journal,* October 13, 74–80.

Sproxton, F. 1925. Cellulose ester varnishes. Review of F. Sproxton, *Cellulose Ester Varnishes. The Decorator,* September 22, 190–91.

Standeven, H. A. L. 2004. The historical and technical development of gloss house-paints, with reference to their use by twentieth-century artists. Ph.D. thesis, Royal College of Art.

Stokes, S. C. 1948. Rubber in flat paints and undercoatings. *The Decorator,* May 15, 66, 93.

Story of Paints Division. n.d. Transcript. ICI Archive, Suffolk Records Office, Ipswich. HCLI 11/5/5 1.

Stoye, D., and W. Freitag. 1998. *Paints, Coatings and Solvents.* 2nd ed. New York: Weinheim, Wiley-VCH.

Stoye, D., W. Freitag, and G. Beuschel. 1996. *Resins for Coatings: Chemistry, Properties and Applications.* New York: Hanser Gardner.

Sutermeister, E., and F. L. Browne, eds. 1939. *Casein and Its Industrial Applications.* New York: Reinhold.

Symons, P. S. 1933. The why and wherefore of modern decoration: Flat varnish. *The Decorator,* July 15, 31–32.

———. 1954. The why and wherefore of modern decoration: The clear oleoresinous finish (part III). *American Paint Journal,* May 17, 78–92.

Synthetic and applied finishes. 1930. *The Decorator,* April 22, 10.

Synthetics. 1950. *The Decorator,* May 15, 41.

Taylor, C. J. A. 1948. Synthetic resins in surface coatings. In *British Catalogue of Plastics: A Plastics Encyclopaedia,* ed. E. Molloy and P. Reilly. London: National Trade Press.

Terminology of paint and painting. 1941. *The Decorator,* September 15, 25.

Terril addresses vehicle manufacturers on castor oil. 1953. *American Paint Journal,* January 26, 14–17.

Thomas Parsons & Sons. n.d. *150 Years of Paint & Varnish Manufacturing.* London: Thos. Parsons & Sons Ltd.

Trade marks. 1931. *Synthetic and Applied Finishes* 1 (12): 171.

Treatment of kitchens. 1947. *The Decorator,* October 15, 38–42.

Troutman, R. E. 1943. Paints and enamels for interiors. In *Protective and Decorative Coatings,* vol. 3, ed. J. Mattiello. New York: John Wiley & Sons; London: Chapman & Hall.

Two new "Dulux" products. 1932. *The Decorator,* August 15, 70.

Uebele, C. E. 1913. *Paint Making and Colour Grinding.* London: Trade Papers Publishing. http://chestofbooks.com/home-improvement/repairs/painting/Paint-Making-Color-Grinding. Accessed September 1, 2009.

United States vs. ICI: Chronological Outline. n.d. DuPont Archive, Hagley Museum and Library, Wilmington, DE.

U.S. National Research Council. 1940. Research: A national resource: II industrial research. December. Available at www.archive.org/stream/researchnational1940unitrich/researchnational1940unitrich_djvu.txt.

The use of boiled linseed oil. 1924. *The Decorator,* May 22, 1001–2.

The uses of rubber in decoration. 1934. *The Decorator,* April 16, 44–47.

Vannoy, W. G. 1943. Exterior finishes. In *Protective and Decorative Coatings,* vol. 3, ed. J. Mattiello. New York: John Wiley & Sons; London: Chapman & Hall.

Varnish gums, resins. 1950. *American Paint Journal,* February 27, 43.

Varnish or lacquer for furniture? 1950. *The Decorator,* April 15, 111.

Vehicle manufacturers hear Davis discuss polyvinyl acetate. 1954. *American Paint Journal,* May 24, 9–10.

Venison, J. R. 1951. Festival paint headache. *The Decorator,* June 15, 50–52.

Ward, G. J. 1926. Brushing lacquers. *The Decorator,* June 22, 947.

Ware, E. E. 1936. Soybean oil and the paint industry. *Industrial Engineering Chemistry* 28 (8): 903–6.

Warson, H. 1972. *The Applications of Synthetic Resin Emulsions.* London: Ernest Benn.

Warson, H., and C. Finch. 2001. *Applications of Synthetic Resin Latices,* vol. 2: *Latices in Surface Coatings: Emulsion Paints.* New York: John Wiley & Sons.

Warth, A. H., and F. L. Browne. 1939. Casein paints. In *Casein and Its Industrial Applications,* ed. E. Sutermeister and F. L. Browne. New York: Reinhold.

Water paint and limewash. 1946. *The Decorator,* July 15, 59.

Wayne, R. 1936. Synthetic resin varnishes. *The Decorator* (January 15): 44.

West, C. J., ed. 1934. *Annual Survey of American Chemistry.* Vol. 3, *1933.* National Research Council. New York: Chemical Catalog Company.

Why good lacquers cost more. 1946. *The Decorator,* June 15, 47.

Why water paint flakes. 1948. *The Decorator,* May 15, 55–56, 93.

Williams, R. 1960. Synthetic latex polymers: A market study. ICI Archive, Catalyst, 7502/1, p. xix.

Wycherley, D. P. 1952. The flaking of water paint. *The Decorator,* November 15, 46–48.

Zintl, W. M. 1947. Departmental histories, 1923–1984: Paints Division. Acc. 1850. DuPont Archive, Hagley Museum and Library, Wilmington, DE.

Index

About the Author

Harriet Standeven is a freelance paintings conservator who specializes in the care of modern and contemporary art. She obtained an MA in painting conservation in 2000 from the University of Northumbria and completed her PhD at the Royal College of Art in 2004. Her research focused on the use of gloss house paints by twentieth-century artists. From 2004 to 2009 she worked as a tutor on the RCA/V&A Conservation Programme. She also conducts research into artists' use of commercial materials.